Path Routing in Mesh Optical Networks

Path Routing in Mesh Optical Networks

Eric Bouillet
IBM Research, USA

Georgios Ellinas
University of Cyprus, Cyprus

Jean-François Labourdette
Verizon Business, USA

Ramu Ramamurthy
Hammerhead Systems, USA

John Wiley & Sons, Ltd

Other Wiley Editorial Offices

John Wiley & Sons Inc., 111 River Street, Hoboken, NJ 07030, USA

Jossey-Bass, 989 Market Street, San Francisco, CA 94103-1741, USA

Wiley-VCH Verlag GmbH, Boschstr. 12, D-69469 Weinheim, Germany

John Wiley & Sons Australia Ltd, 42 McDougall Street, Milton, Queensland 4064, Australia

John Wiley & Sons (Asia) Pte Ltd, 2 Clementi Loop #02-01, Jin Xing Distripark, Singapore 129809

John Wiley & Sons Canada Ltd, 6045 Freemont Blvd, Mississauga, Ontario, L5R 4J3, Canada

Wiley also publishes its books in a variety of electronic formats. Some content that appears
in print may not be available in electronic books.

Library of Congress Cataloging-in-Publication Data:

Path routing in mesh optical networks / Eric Bouillet ... [et al.].
      p. cm.
   ISBN 978-0-470-01565-0 (cloth)
   1. Optical Communications. 2. Routing (Computer network management)   I.
Bouillet, Eric.
   TK5103.59.P38 2007
   621.382′7 – dc22

                                                          2007014321

British Library Cataloguing in Publication Data

A catalogue record for this book is available from the British Library

ISBN: 978-0-470-01565-0 (HB)

Typeset in 9/11pt Times by Laserwords Private Limited, Chennai, India
Printed and bound in Great Britain by Antony Rowe Ltd, Chippenham, Wiltshire
This book is printed on acid-free paper responsibly manufactured from sustainable forestry
in which at least two trees are planted for each one used for paper production.

"To the memory of my beloved grandparents Paraskevas and Theodora" (Georgios Ellinas)

"To my wife Isabelle and my sons Eric and Thomas" (Jean-François Labourdette)

"To my wife Shao-i" (Eric Bouillet)

"To my parents, Lakshmi and Shreyas" (Ramu Ramamurthy)

And to

"All of our Tellium colleagues who made this book possible and contributed to the advancement of mesh optical networks"

Contents

List of Figures

List of Tables

Foreword

The day I first drafted a Foreword to this book one of my graduate students e-mailed me: "Here's *another* one . . ."

December 28, 2006

Asian Quake Disrupts Data Traffic

SEOUL, South Korea, Dec. 27 – Telecommunications across Asia were disrupted on Wednesday after an earthquake off Taiwan damaged undersea cables, jamming Internet services as voice and data traffic vied for space on smaller cables and slower satellite links. The quake disrupted services in Taiwan, Singapore, Hong Kong, South Korea and Japan, but a ripple effect was felt in other parts of the world. Many phone subscribers could not get through to Europe, regional telecommunications operators reported, as they raced to reroute their traffic to alternative lanes. 'We are seeing really massive outages in a spread of countries in East and Southeast Asia,' said Todd Underwood, chief operations and security officer at the Internet monitoring firm Renesys. . . .. (etc.)'

The report goes on to describe how financial companies and businesses in the region were hit hard and how online banking and communications between financial markets and traders were affected.

Why my student said "another one" is much to the point: failures such as this, and even more bizarre and unpredictable ways in which communication networks are disrupted, are an almost daily occurrence. FCC statistics show that metro networks annually experience approximately 13 cuts for every 1000 miles of fiber, and long-haul networks experience 3 cuts for every 1000 miles of fiber. This may sound like a low risk on a per-mile basis, but even the lower rate for long-haul networks implies a cable cut every four days on average in a typical network with 30000 route-miles of fiber. In the first four months of 2002 alone, the FCC logged 50 separate network outages throughout the United States with some very peculiar causes. Of course the most common cause of failure, construction-related dig-ups of fiber-optic cables, is so frequent (despite extensive measures at physical-layer protection) that a wry joke in the industry is to refer to the backhoe as a "Universal Cable Locator". In other recent failures, a fire in a power transformer melted a fiber cable affecting 5000 customers for over 9 hours, a faulty optical amplifier brought down an OC-192 connection between Vancouver and Victoria, BC, for over 10 hours, a boat anchor cut a cable taking 9 days to return an OC-192 connection between Montreal and Halifax back to service, and so on. In fact, in 2004, the whole island nation of Jamaica was disconnected from the world by Hurricane Ivan. First a cable break occurred from wave action in the shallows off Kingston. The country was still connected via a fiber-optic cable to the west through the Cayman Islands. But the Cayman Islands were the next target downrange . . . dead in the sights of Ivan. A redundant cross-island connection there to a major Caribbean regional cable system was then severed, isolating Jamaica from the world for most of a week, aside from some low-capacity satellite connections.

So failures are much more common than most of us would assume or even imagine. And yet backbone fiber-optic transport networks are now absolutely crucial to society. So how do services survive the failure of their underlying physical elements? Certain approaches to the problem are what this book is about and these authors are the sort of "dream team" to write on the topic. Designing and operating networks in a way that services can recover from failures almost instantly is a key aspect of transport networks. Being based in Canada, I can report that here it is seen as so vital and essential an infrastructure that governmental organizations have identified the telecommunications system as one of Canada's ten most critical infrastructures and are keenly funding research to "produce new science-based knowledge and practices to better assess, manage, and mitigate risks to the critical infrastructure." Similar developments and recognition of the telecommunication transport network as critical national infrastructures are well established in the USA and Europe.

This brings us to the timing for this book, and the perfect suitability of these authors to the topic. Bouillet, Ellinas, Labourdette and Ramamurthy were all at Tellium during a phase of history in this field, where remarkable vision, talent, timing and technological mastery combined to breathe real life into one of my own long-held visions, that of distributed mesh-based "Selfhealing networks." Tellium was the provider of the world's first in-service, intelligent optical switch, tested and operated in a nationwide 45-node network by Dynegy's telecommunications subsidiary. The Aurora Optical Switch^{TM} was a 512-port OEO switch of STS-48 (2.5 Gbps) granularity that realized greater network capacity, reliability, and capital efficiencies than network operators had previously seen. Interestingly, during its operating life the Dynegy's network was actually "reoptimized" twice while in service as described in Chapter 10 of the book. Distributed mesh restoration ability was part of the advanced operational capabilities of optical networks built with Tellium cross-connects at a time when some larger vendors were still deadlocked in the ring versus mesh debate. But, at least in my view, the advantages of mesh-oriented operation and survivability were abundantly clear by then and Tellium led the way to practical realization of these potentials. And these authors were at the center of it all. This central Tellium connection between the authors, and their other past experiences mean they write with the authority of technical experience and practical awareness of the issues involved. But the material in the book is more general than just the Tellium experience. An example of another large operational deployment of a distributed mesh-based selfhealing network was the AT&T one using the Ciena CoreDirector^{TM} platform, switching at STS-1 granularity.

Eric Bouillet worked at Tellium on the design of optical networks and optimization of lightpath provisioning and fault restoration algorithms; and before that in the Mathematical Sciences Research Center in Bell Labs/Lucent Technologies on routing and optimizations of telecommunication networks. Georgios Ellinas was a senior network architect at Tellium Inc. In this role, he worked on lightpath provisioning and fault restoration algorithms in optical mesh networks, and the architecture design of another Tellium development project, that of a MEMS-based all-optical (OOO) switch. George also served as a senior research scientist in Telcordia Technologies' (formerly Bellcore) Optical Networking Research Group. Georgios performed research for the Optical Networks Technology Consortium (ONTC), Multiwavelength Optical Networking (MONET) and Next Generation Internet (NGI) projects. Jean-François Labourdette, currently with Verizon Business, was Manager of System Engineering at Tellium, responsible for network element and network management system engineering activities. When he first joined Tellium, he was Manager of Network Routing and Design, responsible for Tellium's routing architecture and algorithms, network dimensioning, and customer network design activities. Previously, he was a Manager of Data Services Globalization Planning at AT&T and before that a System Engineer in the routing planning group, working on dynamic call routing for AT&T's switched network and facility routing and rearrangement for the AT&T transport network. Ramu Ramamurthy has worked in software and systems engineering at Cisco Systems, Ciena Corp, Tellium, Bellcore, and Bay Networks.

Of the only three or four books available to date on the topic of survivable transport network operation and design, I recommend this title as a must-have for network planners, researchers and graduate students in Optical Networking.

Wayne D. Grover, P.Eng, Ph.D, IEEE Fellow, NSERC Steacie Fellow, FEIC
Professor, Department of Electrical and Computer Engineering, University of Alberta
Chief Scientist (Network Systems Research), TRLabs,
Edmonton, Alberta,
Canada

Notes on sources

Interested readers can find information on some of the typical failures cited from CANARIE, "About CA*net 4," available on-line: www.canarie.ca/canet4/index.html, CANARIE, "CA*net 4 Outage Reports." See also D. Crawford, "Fiber Optic Cable Dig-ups: Causes and Cures," Network Reliability: A Report to the Nation–Compendium of Technical Papers, National Engineering Consortium, Chicago, June 1993. FCC Outage reports are also available at Federal Communication Commission. For example, one mentioned is "FCC Outage Report 02–026," FCC Office of Engineering and Technology Outage Reports, February 2002. The Canadian initiative mentioned on critical infrastructure research is the Natural Sciences and Engineering Research Council of Canada, "Joint Infrastructure Interdependencies Research Program (JIIRP)," accessed 8 November 2004, www.nserc.gc.ca/programs/jiirp_e.htm, March 2004. The Dynegy network deployment mentioned in the Foreword is described further in the book's own references [CHAR02], [CHAR03]. The MEMS-based all-optical (OOO) switch development project at Tellium is described in [ELLI03]. AT&T's STS-1 managed Selfhealing mesh network is described in [CORT02] and [RANG02]. (The latter references are referred to in the form they appear in the book's bibliography.)

Preface

People's insatiable appetite and need for communications, trade, entertainment, and access to information, is as old as humanity itself. In today's society, this manifests itself in an increasingly rich set of *network-based* human-to-human, human-to-machine, and machine-to-machine interactions and applications. Examples of such applications include communications through the intermediary of social networking sites such as FaceBook and MySpace. They include B2C trading as exemplified by Amazon.com and now most brick-and-mortar companies, C2C trading with companies such as eBay as well as free-sharing models such as BitTorrent, and B2B trading with supply-chain integration and algorithmic trading. They have expanded into entertainment with massively multiplayer on-line gaming and virtual worlds such as SecondLife. They encompass access to exponentially growing and increasingly accessible information with new paradigms of on-line encyclopedia such as Wikipedia, and search-based access to the huge amount of information available on-line with tools provided by Google, Yahoo, and others. These types of interactions and applications are more than ever served and delivered over the Web and the underlying telecommunications networks. As the supporting information format has evolved from voice and sound to content-rich data and video, that has driven the need for very large amount of flexible bandwidth at the core of the network and all the way to the end-users and end-computers. And this is most certainly only the beginning.

The information and application explosion that we are currently experiencing, is in large part possible due to the radical progress in optical communications technology over the last few decades. Dense Wavelength Division Multiplexed (DWDM)-based optical mesh networks that route optical connections using optical cross-connects (OXCs) have been proposed as the means to implement the next generation optical networks. Optical mesh network architectures as we envision them will dynamically provide transmission capacities to higher-layer networks, such as inter-router connectivity in an increasingly IP-centric service infrastructure. They will also provide the intelligence required for efficient operations, and control and management at the core of the network.

Optical mesh networks will support a variety of dynamic wavelength services, enabling network services such as bandwidth-on-demand, just-in-time bandwidth and bandwidth scheduling, bandwidth brokering, and optical virtual private networks that open up new opportunities for service providers and their customers alike. At the core of this next generation optical mesh network lies the *intelligence* of the optical network elements and network management platforms required to efficiently provide routing and fast failure recovery. That is precisely the subject of this book.

Most of the books on optical communications or optical networks currently available include a host of subjects – from optical transmission technology to general network architectures, planning, analysis, modeling, and management and control. Contrary to that approach, our book presents an in-depth treatment of a specific class of optical networks, namely path-protection oriented mesh optical networks, and focuses specifically on routing and failure recovery associated with Dedicated Backup Path Protection (DBPP) and Shared Backup Path Protection (SBPP). This book focuses on the routing, recovery, dimensioning, performance analysis, and availability in such networks. This book is intended as a reference for practicing engineers working on the deployment of intelligent fiber-optic networks, and for researchers investigating a host of problems on this subject. This book is not meant for readers interested in fiber-optic communications in general, as it does not provide information about optical transmission at the physical layer, or the technology required for the deployment of such an intelligent optical network. There are a large number of such books in the literature, including, for example, Agrawal[1], Keiser[2] and Palais[3] for the reader seeking a deeper understanding of the underlying optical components and transmission technology, as well as a number of general texts on optical networks (such as books by Stern, Ellinas and Bala,[4] by Ramaswami and Sivarajan,[5] by Jukan[6] and Mukherjee.[7] In addition, readers who seek a more detailed understanding on the control plane of optical networks have resources such as the book by Bernstein, Rajagopalan and Saha[8] and readers who want a deep understanding of survivability in optical as well as MPLS, SONET and ATM networks are encouraged to read such references as the books by Grover,[9] Mouftah and Ho[10] and Zang.[11]

Different parts of the book will be appropriate for different audiences. Some chapters will be of more interest to network planners and designers, while others, more forward-looking, will be of more interest to researchers.

Chapters 1, 2, and 3 are suitable for a reader who wants to gain some qualitative knowledge of intelligent optical networks. These chapters give a basic description of mesh optical networks and the basic concepts on routing and restoration in such networks, without treating these subjects in depth. Chapter 1 explains the evolution of optical networks and discusses different network architectures, Chapter 2 describes the numerous survivability techniques that are available for optical networks in general, and Chapter 3 focuses on routing and survivability concepts for shared mesh optical networks in particular.

Detailed discussions on algorithms for routing and Dedicated and Shared Backup Path Protection in mesh optical networks are presented in Chapters 4 through 10 and this material will be of interest to practicing engineers and researchers who are currently deploying or investigating the benefits of intelligent mesh optical networks. Chapter 4 introduces and focuses on the specific routing and recovery framework covered and studied in the reminder of the book: mesh optical networks operated with a path-based protection architecture, in particular Dedicated Backup Path Protection (DBPP) and Shared Backup Path Protection (SBPP). Chapter 5 presents a detailed introduction to and discussion of the algorithmic aspects of routing in path-protected mesh networks, and assesses the corresponding routing complexity. Chapter 6 discusses a number of practical and efficient routing heuristics, Chapter 7 describes advanced cost metrics that can be incorporated in these heuristics to drive certain network behaviors, and Chapter 8 describes ways of controlling and managing the amount of sharing through additional modifications of these heuristics. Chapter 9 takes the problem of routing and recovery in mesh optical networks a step further by investigating techniques for route

---

[1]G. P. Agrawal, *Fiber-Optic Communication Systems,* Third Edition (Wiley Series in Microwave and Optical Engineering). New York: John Wiley & Sons, 2002.

[2]G. Keiser, *Optical Fiber Communications,* Third Edition. McGraw-Hill, 2000.

[3]J. C. Palais, *Fiber-Optic Communications,* Fifth Edition. New Jersey: Prentice Hall, 2004.

[4]T. E. Stern, G. Ellinas and K. Bala. *Multiwavelength Optical Networks: Architectural Design and Control,* 2nd Edition, Cambridge University Press, 2007.

[5]R. Ramaswami and K. Sivarajan, *Optical Networks: A Practical Perspective,* Second Edition. Morgan Kaufmann, 2002.

[6]A. Jukan, *QoS-based Wavelength Routing in Multi-Service WDM Networks.* Springer, 2001.

[7]B. Mukherjee, *Optical WDM Networks.* Springer, 2006.

[8]G. Bernstein, B. Rajagopalan and D. Saha, *Optical Network Control.* Addison-Wesley, 2004.

[9]W. Grover, *Mesh-Based Survivable Networks.* New Jersey: Prentice Hall, 2004.

[10]H. Mouftah and P-H. Ho, *Optical Networks: Architecture and Survivability.* Kluwer Academic Publishers, 2003.

[11]H. Zang, *WDM Mesh Networks: Management and Survivability.* Kluwer Academic Publishers, 2003.

computation with partial network information. In Chapter 10, we address the problem of reoptimizing the network and rerouting of connections over time as demand changes and the network infrastructure evolves.

Finally, Chapter 11 addresses the dimensioning and recovery performance of mesh optical networks through analytical means, while Chapter 12 covers and studies the service availability of path-protected connections in these networks. These chapters will be of interest to engineers who are interested in the dimensioning and capacity planning aspects of mesh optical networks, and to those who want to understand the availability performance that can be achieved for path-protected connections, and how it relates, or not, to recovery times. This work is also relevant to researchers, as routing with availability objectives or constraints, and network dimensioning, are two subjects that are actively being investigated by researchers in the area of optical networking. The reader is referred to *http://www.eng.ucy.ac.cy/gellinas/book.html* for useful resources such as a web service to apply the mesh routing algorithms described in the book as well as several case studies.

Acknowledgments. Most of the work presented in this book either originated or was performed when all the authors were part of the Network Routing & Design Group under Jean-François Labourdette, at Tellium Inc. This group was responsible for the development of routing and protection/restoration architectures and algorithms, the design and performance modeling and analysis of mesh optical networks, including comparison with other network and service architectures, as well as numerous customer studies.

We are indebted to our colleagues Sid Chaudhuri and Krishna Bala, the immediate supervisors of the group, for their support, encouragement and expert guidance during the Tellium years. Their broad knowledge of telecommunications networks in general, and of optical networks in particular, was a tremendous resource for us. Their acute understanding and vision gave us valuable insight into several aspects of optical networking and helped us widen our perspectives.

In Tellium, we also had the privilege of interacting with an excellent group of colleagues who were instrumental in the development of the material presented in this book. In particular, we would like to acknowledge the other two members of the group, Ahmet Akyamaç and Chris Olzewski. Ahmet Akyamaç was instrumental in the work on modeling of mesh optical networks and Chris Olzewski in the work on multi-tier networks. Many fruitful discussions with them provided us with probing criticism, valuable suggestions and comments. It was a pleasure and a privilege working with them both intellectually and personally.

Special thanks also go to a number of other colleagues at Tellium, including Subir Biswas, Ziggy Bogdanomicz, Matt Busche, Samir Datta, Somdip Datta, Evan Goldstein, Nooshin Komaee, Lih Lin, Hang Liu, Dimitris Pendarakis, Bala Rajagopalan, Debanjan Saha, Shahrokh Samieian, Sudipta Sengupta, Bo Tang, and James Walker for the many enlightening discussions on endless technical subjects that have greatly benefited this work.

The quality of this book has significantly improved due to the diligent efforts of a number of unnamed colleagues who took the time to read it and provide valuable comments and suggestions. Our many thanks go out to them. We would also be remiss if we did not extend a thank you to Professor Wayne Grover for taking the time to review early drafts of this book and provide incredibly valuable feedback, and for writing the Foreword for this book.

We also wish to express our thanks to Richard Davies, Birgit Gruber, Sarah Hinton, Rowan January, Joanna Tootill, and Brett Wells, the publishing team at Wiley, for their diligence, and patience while dealing with missed deadlines and last minute changes.

Finally, Georgios Ellinas is greatly indebted to his mother Mary and sister Dorita for their faith and confidence in him, their unyielding emotional support, understanding and patience that have been a source of infinite strength during this undertaking. Jean-François Labourdette wishes to express his profound gratitude to his wife Isabelle and his two children Eric and Thomas for their support throughout this endeavor and for their understanding when he could not spend as much time with them as he would have liked. Eric Bouillet is most indebted to his parents and brothers for their encouragement and support, and to his wonderful wife Shao-i for her love, patience, and the comfort of her unwavering kindness. Ramu Ramamurthy thanks his wife Lakshmi, and his parents for their constant encouragement and support, and his son Shreyas for providing the necessary distraction.

Eric Bouillet
Georgios Ellinas
Jean-François Labourdette
Ramu Ramamurthy

Chapter 1

Optical Networking

1.1 Evolution of Optical Network Architectures

The size and complexity of telecommunications networks and the speed of information exchange have increased at an unprecedented rate over the last decades. We live in a new information era, where most people are currently using a number of devices with advanced multimedia applications to obtain and exchange information. The current trends in multimedia communications include voice, video, data and images. These trends are creating a demand for flexible networks with extremely high capacities that can accommodate the expected vast growth in the network traffic volume.

In today's integrated networks, a single communications medium should be able to handle individual sessions with a variety of characteristics, operating in the range of a few megabits to tens of gigabits per second. This will enable it to handle such applications as large-volume data or image transfers (e.g., supercomputer interconnections, supercomputer visualization and high-resolution uncompressed medical images) that have very large bandwidth requirements, as well as applications such as voice or video which require much smaller bandwidth.

The enormous potential of optical fiber to satisfy the demand for these networks has been well established over the last few decades. Optical fiber is highly reliable (Bit Error Rate (BER) in commercially deployed systems is less than 10^{-12}), it can accommodate longer repeater spacings and it has unlimited growth potential. Single mode fiber offers a transmission medium with Tbps bandwidth (enough capacity to deliver a channel of 100 Mbps to hundreds of thousands of users) combined with low loss and low BER. Traditional network architectures, however, that used electrical switches and the optical fiber as a simple substitute for copper wire or other communications media, were limited by an electronic speed bottleneck and could not have been used in telecommunications networks with a growing demand for Gbps applications.

As the next step in the evolution of transport networks, Wavelength Division Multiplexed (WDM) optical networks were proposed [58] which provided concurrency by multiplexing a number of wavelengths for simultaneous transmission within the same medium. Rapid advances in optical fiber communications technology and devices, in terms of performance, reliability and cost over the last few years, were the catalyst in enabling the deployment of optically routed WDM networks as the next generation, high-capacity nationwide broadband networks [289]. This approach can then provide each user with a manageable portion of the enormous aggregate bandwidth.

Path Routing in Mesh Optical Networks   Eric Bouillet, Georgios Ellinas,
Jean-François Labourdette, Ramu Ramamurthy   © 2007 John Wiley & Sons, Ltd

The current information explosion is indeed in large part due to the radical progress in optical communications technology over the last few decades. Dense Wavelength Division Multiplexed (DWDM) mesh networks that route optical connections using optical cross-connects (OXCs) have been proposed as the means to implement the next generation optical networks [308]. Following a wave of timely technological breakthroughs, optical network equipment vendors have developed a variety of optical switching systems capable of exchanging and redirecting several terabits of information per second. The dimensions of the switches range from a few tens to several hundred ports with each single port capable of carrying millions of voice calls, or thousands of video streams. The emergence of new optical technologies is driving down the overall network cost per units of bandwidth, and the trend is accompanied by an explosion of new data service types with various bandwidth characteristics and prescribed Quality of Service (QoS). Optical network architectures as we envision them now not only provide transmission capacities to higher transport levels, such as inter-router connectivity in an Internet Protocol (IP)-centric infrastructure, but also provide the intelligence required for efficient routing and fast failure recovery in core networks [36, 212, 304]. This is possible due to the emergence of optical network elements that have the intelligence required to efficiently manage such networks. Figure 1.1 illustrates the optical network hierarchy, with a core optical network incorporating mesh topologies with optical cross-connects interconnecting WDM metro networks incorporating reconfigurable optical add drop multiplexers (ROADMs), which in turn interconnect various access networks.

The reader should note that there is another alternative architecture in which the IP routers are directly connected to WDM systems (i.e., there is no optical switching). Historically, there have existed two schools of thought concerning the evolution of the core network architecture. The first argued that all of the intelligence should reside within the IP layer, and the optical layer should just be used for transport, while the second argued to move away from a network where all the processing is done at the IP layer to a network where the intelligence is shared between the IP routers and the optical cross-connects. Figure 1.2 illustrates the two different network architecture scenarios. We believe the latter vision is more appropriate for core networks for a variety of reasons. For example, for

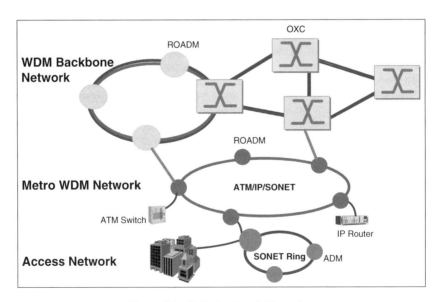

Figure 1.1: Optical network hierarchy.

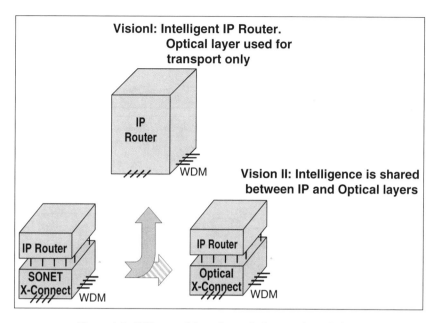

Figure 1.2: Different visions for optical network evolution.

the case of failure recovery that is a main focus of this book, protection/restoration in the optical layer is typically faster, more robust and simpler to plan and upgrade compared to IP/Multiprotocol Label Switching (MPLS)-based recovery. Thus, even though the first vision may indeed be a viable architecture it is not discussed in the remainder of this book, which deals exclusively with the case where intelligent optical switching is present in the network.

Optical networks enable a variety of wavelength services (such as wavelength-on-demand, wavelength brokering, and optical virtual private networks) that open up new opportunities for service providers and their customers alike. In addition to new services, high-speed connections at 2.5 Gbps rates and above are required for the optical core to support trunking between edge service platforms. The dominant traffic carried in today's network is evolving from legacy voice and leased line services to data services, predominantly IP services. Time Division Multiplexing (TDM) aggregation switches, optimized for legacy voice services and leased line services, and acting as edge devices, groom signals at lower bit rates (e.g., 1.728 Mbps and 51.840 Mbps), and feed them into the core, typically, at rates of 2.5 Gbps and 10 Gbps. Equipment operating at 2.5 and 10 Gbps are currently commercially available, and there has been considerable work (both experimentally and in some cases some initial commercial products were developed) on 40 Gbps data rate transmission and switching[1] [47, 69, 70, 150].

Figure 1.3 illustrates the four different node architectures that can comprise a reconfigurable core optical network. The first architecture shows a fixed patch panel. Fixed patch panels located between WDM systems with transponders are currently being replaced by opaque switching nodes (with electrical switch fabrics), as shown in the architecture of Figure 1.3(b), due to their complete lack of flexibility. This is an *opaque network architecture*, as the optical signal now undergoes Optical–Electrical–Optical (OEO) conversion at the switch [30]. The third architecture shows a transparent switch between WDM systems with transponders that would be complemented by an OEO switch for drop traffic. This is again an opaque network architecture, as the optical signal

---

[1]Commercial deployments are also taking place.

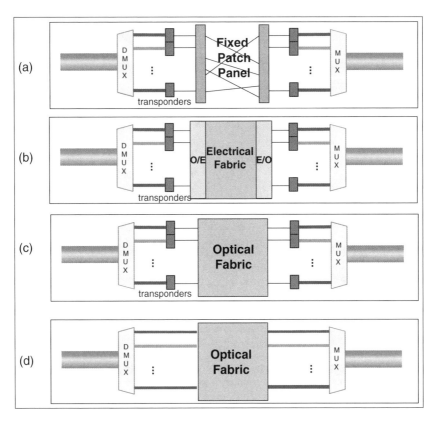

Figure 1.3: Node architectures for a core optical network. (a) Opaque network with fixed patch panel, (b) Opaque network with opaque switch, (c) Opaque network with a transparent switch and (d) Transparent network with a transparent switch. (After [108], Figure 1. Reproduced by permission of © 2003 The International Engineering Consortium.)

undergoes OEO conversion at the WDM transponders. The fourth architecture shows a completely *transparent* network topology, consisting of transparent optical switches and WDM systems that contain no transponders. The transparent switch would be complemented as in Figure 1.3(c) by an OEO switch for drop traffic. In this architecture, the signal stays in the optical domain until it exits the network. Details on the design of each architecture are presented in the sections that follow. There has been extensive research work on the limits of optical transparency, comparisons between transparent and opaque networks, and the benefits and drawbacks of each technique. The reader is referred to [203, 268, 299] for some of the work that has been performed in that area.

1.1.1 Transparent Networks

The transparent node architecture shown in Figure 1.3(d) and elaborated on in Figure 1.4 is a seemingly attractive vision. A signal (wavelength) passing through an office does not undergo opto-electronic (O/E) conversion. Similarly, a client Network Element (NE), such as a router, interfaces with the switch using long-haul optics to interface with the WDM equipment without any O/E conversion. Since a signal from a client NE connected via a specific wavelength must remain on the same wavelength when there is no wavelength conversion, only a small-size switch fabric is needed to

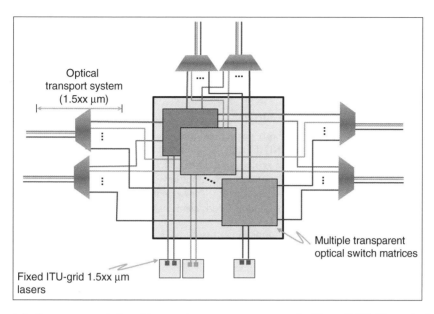

Figure 1.4: Transparent switch architecture in a transparent network. (From [109], Figure 1. Reproduced by permission of © 2004 The Institute of Electrical and Electronics Engineers.)

interconnect the WDMs and NEs in a node. This architecture also implies end-to-end bit rate and data format transparency. Note that another architecture of a transparent switch in a transparent network may include a single large fabric instead of multiple switch matrices of small port counts. However, if one is to provide flexibility, such an architecture design would require the use of tunable lasers at the clients and wavelength conversion.

A transparent network architecture may provide significant footprint and power savings and on the surface suggests cost savings. However, while the transparent network architecture may be a viable option for small-scale networks with pre-determined routes and limited numbers of nodes, it may not be a practical solution for a core mesh optical network for the following reasons:

- This network does not allow wavelength conversion, thus essentially creating a network of n (n being the number of WDM channels) disjoint layers. Inflexible usage of wavelengths in this network would lead to increased bandwidth and network operational cost, thus negating all savings that may result from the elimination of O/E conversion. In addition, for this technology to be effective and in order to build a flexible network for unrestricted routing and redundant capacity sharing, an all-optical 3R-regeneration function must be available. Such a technology that can be harnessed in a commercial product does not currently exist [220].

- In the absence of wavelength conversion, only client-based dedicated backup path protection (DBPP) can be easily provided [107, 187]. The wavelength continuity constraint on backup paths makes resource sharing very difficult in transparent networks and consequently no shared backup path protection (SBPP)[2] can be easily offered. This in turn means that the capacity requirement for protected services is significantly higher (80–100%) for transparent compared to opaque networks [68].

---

[2]The concepts of sharing, DBPP and SBPP, will be explained in detail in Chapter 2. SBPP is sometimes alternatively termed *backup multiplexed* path protection.

- Physical impairments such as chromatic dispersion, polarization mode dispersion (PMD), fiber nonlinearities, polarization-dependent degradations, WDM filter pass-band narrowing, component crosstalk, amplifier noise, etc. accumulate over the physical path of the signal due to the absence of O/E conversion. The accumulation of these impairments requires engineering of end-to-end systems in fixed configurations [197, 245, 246, 247]. Thus, it may not be possible to build a large network with an acceptable degree of flexibility.

- The design of high-capacity DWDM systems is based on intricate proprietary techniques, eluding any hope of interoperability among multiple vendors in the foreseeable future. Since a signal is launched at the client NE through the all-optical switch directly into the WDM system without O/E conversion, and it is not possible to develop a standard for the interface for a high capacity WDM system, the operators will not have the flexibility to select the client NE vendor and the WDM vendor independently. Consequently, transparent networks by necessity are single vendor (including the client network elements) solutions.

- Finally, in addition to all the limitations discussed above, the challenge of performance-engineering continental-scale transparent reconfigurable wavelength-routed networks remains severe and, in networks that push limits, remains unsolved despite some attempts at formalizing the routing problem [290].

It is apparent that a number of key carrier requirements – dynamic configuration, wavelength conversion, multi-vendor interoperability of transport equipment (WDM), low network-level cost – would be very hard to meet in a transparent network architecture. Therefore, an opaque network solution will remain for now the only practical and cost-effective way of building a dynamic, scalable, and manageable core backbone network. A description of the opaque network architecture is offered in the section that follows.

1.1.2   Opaque Networks

Even though the opaque network solution may be more expensive in terms of equipment costs when the core network capacity increases significantly, the opaque network offers the following key ingredients for a large-scale manageable network:

- No cascading of physical impairments. This eliminates the need to engineer end-to-end systems (only span engineering is required) and allows full flexibility in signal routing.

- Multi-vendor interoperability using standard intra-office interfaces (see Figure 1.5).

- Wavelength conversion enabled. Network capacity can be utilized for service without any restrictions and additional significant cost savings can be offered by *sharing* redundant capacity in a mesh architecture (see Figure 1.5).

- Use of an all-optical switch fabric without any compromise of the control and management functions. Overhead visibility (available through an OEO function that could be complementing the all-optical switch) provides support for the management and control functions that are taken for granted in today's networks.

- The network size and the length of the lightpaths can be large, since regeneration and retiming are present along the physical path of the signal.

- Link-by-link network evolution. Permits link-by-link incorporation of new technology, as the network is partitioned into point-to-point optical links (see Figure 1.6).

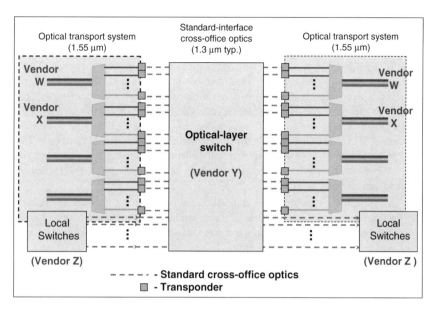

Figure 1.5: Multi-vendor interoperability and wavelength translation as a by-product of an opaque network architecture. (From [109], Figure 2. Reproduced by permission of © 2004 The Institute of Electrical and Electronics Engineers.)

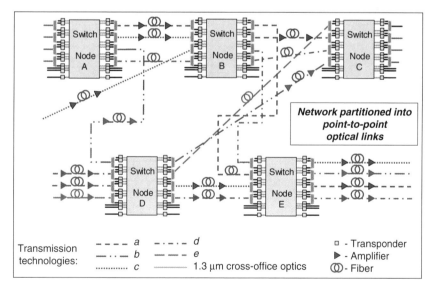

Figure 1.6: Link-by-link network evolution.

Having reasoned that transparent core mesh network architectures are likely to remain unrealistic for quite some time, we now turn our attention to opaque network architectures in which WDM systems utilize transponders. Today's architectures contain, in the most part, opaque switches (with an electronic switch fabric) in an opaque network (with transponders present in the WDM system). This architecture is shown in Figure 1.7. The interfaces to the fabric are opaque interfaces, which means that transceivers are present at all interfaces to the switch, and these transceivers provide an OE (input) and EO (output) conversion of the signal. The presence of the transceivers at the edges of the switch fabric enables the switch to access the Synchronous Optical NETwork/Synchronous Digital Hierarchy (SONET/SDH) overhead bytes for control and management functions. The opaque transceivers provide support for fault detection and isolation, performance monitoring, connection verification, neighbor/topology discovery and signaling, as well as support for implementing the network routing and recovery protocols.

Another design of the opaque switch architecture may be one using Photonic Integrated Circuits (PICs) [222] which can include large numbers of lasers, modulators, and optical multiplexers or optical demultiplexers and photodiodes integrated on the same monolithic Indium Phosphate (InP) chip. This approach allows for low-cost opaque architectures by enabling low-cost OEO conversion of the signal on a semiconductor chip, and is a leap in technology.

The opaque switch approach was, however, faced with a number of challenges when confronted with the traffic growth projections from just a few years ago: it would eventually reach scaling limitations in signal bit rate, switch matrix port count, and network element cost. These were the key motivations behind the attempt to develop large port-count transparent switches to be used in opaque network architectures. For high port-count fabrics, analog gimbal-mirror MEMS (Micro-Electro-Mechanical Systems)-based switches (3D switches) offer the most viable approach [121, 310]. It is important to point out that the opaque switches could still remain in the network architecture in order

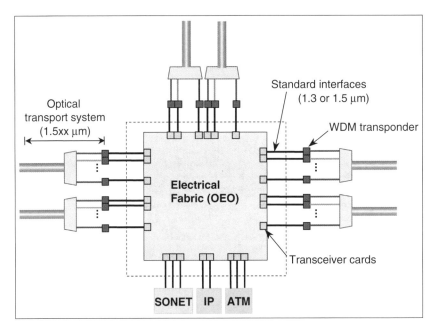

Figure 1.7: Opaque switch architecture. (From [109], Figure 3. Reproduced by permission of © 2004 The Institute of Electrical and Electronics Engineers.)

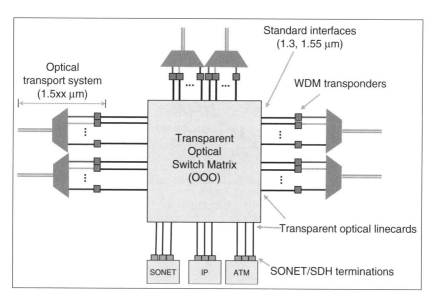

Figure 1.8: Transparent switch architecture. (From [109], Figure 3. Reproduced by permission of © 2004 The Institute of Electrical and Electronics Engineers.)

to provide some key network functions, namely grooming and multiplexing, Service Level Agreement (SLA) verification, and control and management.

Figure 1.8 shows a transparent (OOO) switch architecture. In this architecture, optical signals pass through e.g., a MEMS-based switch fabric, in contrast to the OEO architectures where switching is accomplished using an electrical switch fabric. This switch architecture has transparent interface cards, i.e. no (OEO) transceiver (TR) cards are located at the network ports of the switch fabric that convert the optical into an electrical signal. The switch shown in Figure 1.8 also has no opaque transceiver cards on its add/drop ports. Therefore, it has no direct access to the overhead bytes for control and signaling. The optical switch fabric is bit-rate independent and it accommodates any data rates available (e.g., 2.5, 10 and 40 Gbps).

The advantage of such a switch architecture is that for an $N \times N$ all-optical (OOO) architecture there are N interfaces/ports to the all-optical switch regardless of the type of interfaces. No data-rate-specific interface cards are used, so no replacement is needed when the switch is operating at higher data rates, provided that the optical power budget is sufficient for that rate. This is in contrast to the OEO systems where the number of ports depends on the type of the port. For example, in an OEO system one 10 Gbps interface card will replace four 2.5 Gbps interface cards. From the interface card perspective all ports in the OOO architecture will look one and the same (the same port cards are used for different signal rates and formats). The add/drop-side ports are connected to an OEO switch (or any other client – such as IP/MPLS router or Asynchronous Transfer Mode (ATM) switch) that provides SONET/SDH termination through its opaque ports.

The promise of optical switching was that, unlike integrated electronic switches, an optical switch fabric's complexity is a flat function, independent of the bit rate of the signals it handles (Figure 1.9). Moreover, in the long run, it was projected that few components would be as small, cheap, and low in power consumption as a silicon micro-mirror in the case of MEMS-based switch fabrics. Transparent switches could thus be expected to be cheaper in terms of the switching fabric and interface card cost compared to opaque switches. This would have resulted in significant cost reduction to network

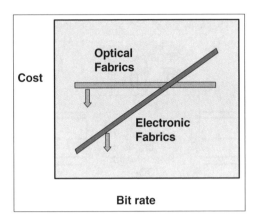

Figure 1.9: Advantages of optical fabrics. (After [108], Figure 5. Reproduced by permission of ©️ 2003 The International Engineering Consortium.)

operators because a large amount of the traffic that passes through an office will be able to bypass the OEO switch (typically approximately 75% through-to-total ratio).

Transparent switches essentially would have helped to relieve the demand for OEO switch ports and reduce the cost of transporting lightpaths. This is accomplished by having all lightpaths pass through the OOO switches (glass through), thus bypassing the OEO switches. Note that this can be a significant portion of the network traffic. For example, if 40 Gbps data rates were used, every time a lightpath passes through an OOO switch, 32 equivalent 2.5 Gbps ports of an OEO switch would be saved (two 40 Gbps ports that correspond to 32 equivalent 2.5 Gbps ports).

Since the OOO switch fabric is bit-rate and data-format independent, the switch matrix can scale more easily than electrical switch fabrics. For these reasons, as bit rates rise, it was thought that optical switch fabrics would eventually prevail. Note, however, that this evolution would only have happened on timescales that were gated by the ability of vendors to meet carrier reliability and operational requirements with all-optical technologies such as lightwave micro-machine (for MEMS-based switch fabric) technology [121]. Even though in early stages of 2.5 Gbps and 10 Gbps development the crossover point shown in Figure 1.9 appeared to be at the 2.5 Gbps and then the 10 Gbps rates, under today's more realistic traffic growth scenario, and given the lack of deployment of 40 Gbps WDM systems and the continued decline in price of OEO components, the crossover point has shifted to the even higher bit rates. Therefore, the need for and the promise of transparent switches appear to have moved beyond the foreseeable future. Provided that the traffic grows and the bit rates increase substantially, there may emerge in the future a potential need for an additional network layer utilizing transparent optical switches. In that case, the main challenge to architectures that use transparent switches will then be to provide the control and management functionalities that are readily available when we have access to the electrical signal and consequently to the SONET/SDH overhead bytes.

Even though the use, in the core, of transparent switches that are cost-effective at very high bit rates is not currently justified, there still exist some niche applications in today's networks that could use a small number of transparent switches. Transparency is mainly limited in metropolitan area networks, utilizing ROADMs, and some ultra long-haul applications in the core, utilizing a small number of wavelength-selective cross-connects/OADMs on high-capacity routes. When OADMs are utilized, selected wavelengths can be added or dropped at a node while the rest of the wavelengths pass through without regeneration [29, 30]. ROADMs further enable any user to access any channel.

ROADMs can be utilized in metropolitan area networks at central offices and customer locations in much the same way that the SONET introduction created a need for large numbers of SONET Add Drop Multiplexers (ADMs). They provide network flexibility and can be used to manage continuous changing traffic patterns and customer service requirements.

Wavelength-selective cross-connects (WSXCs) can also be used in ultra long-haul applications in the core network in a completely transparent manner. Even though these network elements allow for end-to-end bit rate and data format transparency, they face a number of challenges. However, these network elements could be utilized in a few, predetermined and non-reconfigurable high-capacity routes to provide end-to-end transparency between fixed end-nodes. Furthermore, we anticipate that opaque switches will always remain for the embedded service base even after the transparent switches are introduced in the network. These opaque switches will provide the grooming and multiplexing functions, as well as some of the necessary control and management functions, and will scale and decrease in cost with rapid progress in electronics.

While completely transparent core mesh networks have not yet materialized on a large scale, even transparent switches in opaque networks still face technological as well as control and management challenges [109]. Even though most of these issues can be addressed via clever innovation as well as standardization efforts, transparent switches complemented by an opaque function will not be ready for deployment in the network until all the control and management challenges are successfully resolved.

1.1.3 Translucent Networks

There is a third network architecture worth mentioning, the *translucent* network architecture, which is based on optical cross-connects that are a mixture of opaque and transparent cross-connects presented in the previous sections [247, 248]. Figure 1.10 shows an example of such a node. This node is composed of two parts: a transparent and an opaque part. A signal entering this node can pass through (transparently), or can be dropped (or added) and go through a regeneration process [218]. Long light-paths that cause the optical signal to degrade are the candidates that will go through the regeneration process to recover the original signal. Signal regeneration may occur a number of times before the signal reaches its final destination. Translucent networks allow the connections to stay transparent for as long as their signal quality allows, and then go through a regeneration process. These networks thus have some of the advantages of the transparent networks discussed previously while mitigating some of the drawbacks that appear in networks that are completely transparent (such as allowing

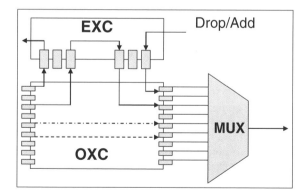

Figure 1.10: Translucent node architecture. (After [218], Figure 1.b. Reproduced by permission of © 2005 DRCN.)

for wavelength conversion and addressing the problem of accumulation of physical impairments on the path).

Potentially not all the nodes in the network will have the regeneration capability. Sparse regeneration can be offered by placing the translucent nodes at strategic locations. Problems such as recognizing and addressing regeneration demands [248], dynamic routing in translucent optical networks [327], and placement of the translucent nodes in order to minimize blocking in these networks [277] have been addressed by the research community.[3]

Studies presented in [217, 218] address the impact of the reach of the WDM systems and the impact of the traffic volume on the cost of these networks. These studies also include cost comparisons between translucent and opaque network architectures. Even though the initial studies presented in [217] showed the translucent architecture to be 50% cheaper than the opaque one, subsequent studies presented in [218] reduced that savings number 10%.[4] The authors in [218] argue that considerable increase in traffic or considerable reduction in the cost of transparent devices will be the factors that can make the translucent network option cost-effective.

1.2   Layered Network Architecture

In this section we review the fundamental parts that constitute a network and its functionality. It goes without saying that many architectures exist or have been suggested and it is not the intention of this book to enumerate them exhaustively (see [39, 155, 243, 270] for further information and useful references on this topic). However, we observe that all the proposed architectures repose on a common denominator. It is this generic model that we present here. The model consists of three superimposed layers. Each layer provides well-defined services to its superjacent layer while concealing implementation details from it.

Figure 1.11 shows a layered architecture with the DWDM network being used as the transport network. The fiber-optic links and optical switching nodes are located in the physical topology. The logical layer represents the view of the network seen by end users, accessing the physical layer through electro-optic interfaces. The access means may be direct (through clear channels) or indirect

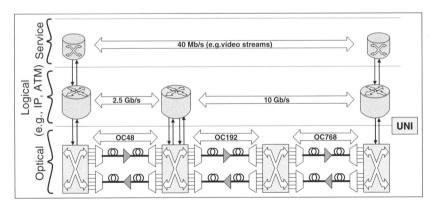

Figure 1.11: Layered network architecture. (From [107], Figure 1. Reproduced by permission of © 2003 The International Society for Optical Engineering.)

---

[3]Apart from the indicative references given here the reader is encouraged to investigate the large body of work that exists on translucent networks, sparse wavelength assignment, etc.

[4]Savings depend in a large part on the architectures of the nodes in the network and on the system reach.

through the intermediary of electronic (e.g., SONET, ATM) switching equipment. The service layer demonstrates the large number of applications that can be provided in such networks. As shown in Figure 1.11, from bottom to top the layers are (1) optical layer, (2) logical (electrical) layer and (3) service/application layer.

1.2.1 Optical Layer

The optical layer offers and manages the capacity required to transport traffic between clients in the logical layer. The optical layer includes wavelength transmission equipment (DWDM), wavelength switching or cross-connect equipment (also called optical switches) handling 2.5 and 10 Gbps wavelengths, and wavelength grooming equipment, handling subrate circuits (in multiples of STS-1) into 2.5 Gbps and/or 10 Gbps wavelengths.

Figure 1.12 depicts an example of a logical network (two IP/MPLS routers) linked to an optical network (four optical switches). Optical switch ports are either: (1) add/drop-ports, interfacing the optical layer to the client's logical layer, or (2) network ports, interconnecting optical switches. Using our graph representation, nodes are optical switches, and links are bundles of bidirectional optical channels between pairs of optical switches. An optical channel is a wavelength that connects the network ports of adjacent optical switches. A link in the logical layer is realized by way of optical channels in tandem forming a lightpath (circuit) between the end-nodes of that link.

The optical layer faces the same challenges, and conceptually even borrows solutions from the logical layer. For instance, it relies on Generalized MPLS (GMPLS) [26, 182, 210] also formerly known as MPLambdaS (an extension of MPLS) to encompass all types of architectures, including wavelength-oriented traffic engineering and management. It also relies on Neighbor Discovery Protocol (NDP)[5]/Link Management Protocol (LMP) [193] and Open Shortest Path First (OSPF) protocol [219] together with Link State Advertisements (LSAs) exchanges, to create and publicize the network's topological views. Differences that set apart the optical layer from its logical counterpart

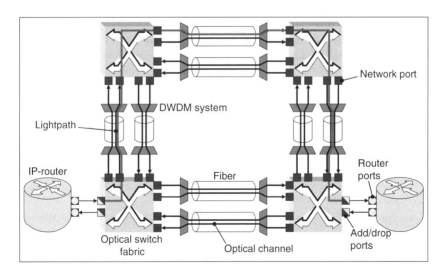

Figure 1.12: The optical layer. (From [107], Figure 2. Reproduced by permission of © 2003 The International Society for Optical Engineering.)

---

[5]The Hello Protocol is also used.

are among others: (1) routing in the optical layer is exclusively circuit oriented, (2) circuit set-up and tear-down is done at a much slower timescale and (3) the bandwidth granularity of the logical layer is much lower than the granularity of the optical layer.

In the *overlay approach* assumed throughout this book the layers work individually, with the client logical layer leasing resources from the optical layer. The User Network Interface (UNI) harmonizes communication of control messages between the two domains [12]. In addition, since an optical carrier will normally acquire network components from several vendors, a suite of protocols is being developed in the Internet Engineering Task Force (IETF) to allow for the seamless interaction between the various network components. As part of that suite, the Link Management Protocol, for example, is used to maintain control channel connectivity, verify component link connectivity and isolate link, fiber or channel failures within the network [12].

1.2.2   Logical Layer

Also known as electrical or digital layer, the logical layer aggregates services into large *transmission pipes* and assures their proper routing from Point of Presence (PoP) to PoP with prescribed QoS [172, 173]. Using a graph representation, a logical node corresponds, for example, to an IP/MPLS core router, an ATM backbone switch or a digital cross-connect (DCS), and a logical link connects the ports of two adjacent nodes. The logical layer may consist of several interconnected subnetworks, either for scalability reasons, as it is easier to manage several smaller networks than a large network (hierarchical decomposition), or because the subnetworks belong to several independent carriers or employ different technologies (e.g., IP/MPLS versus ATM). In either case, boundaries and proper network interfaces within the logical layer delimit the subnetworks and their respective domains of operation.

The logical layer fulfils several roles: (1) it maintains a consistent topological view of its layer, (2) it manages the address space, (3) it routes streams on request, and (4) it polices the traffic to ensure a fair share of capacity among data streams and to guarantee each individual's QoS. The first part, also called topology discovery, can be achieved, for example, by way of an NDP in conjunction with the OSPF protocol [219]. NDP operates in a distributed manner through in-band signaling to construct local port-to-port connectivity databases at each node. OSPF completes the topology discovery by assembling and globally disseminating pieces of information collected by NDP, plus additional information such as link states, to the logical plane [38]. Logical nodes have only a few tens of ports, and with the exception of very small networks, a full connectivity featuring one link between every pair of node is not probable. Instead, services may have to be routed in the logical layer through one or more transit nodes to the desired destination using, for example, Constrained-based Routing Label Distribution Protocol/Resource Reservation Protocol (CR-LDP/RSVP) explicit routing and bandwidth reservation protocols [27]. The computation of a logical path must satisfy a set of constraints, such as round-trip delays and spare bandwidth, defined in the service layer in accordance to prescribed QoS [173]. Note that the failure of a logical link or logical node can be detected, for example, by NDP, and advertised by OSPF. That is, the layer has the primitives to detect a failure and resume interrupted services.

1.2.3   Service/Application Layer

In the service layer, clients such as edge or service routers or Multi-Service Provisioning Platforms (MSPPs) located in a provider's POP represent users and the data communication among them. Using a graph representation, a node corresponds to a client who emits and receives data, and a link represents a service or a two-way data stream between clients. Link attributes in this layer correspond to minimum QoS requirements, which transpose into bit rates, jitter, and bit-propagation or round-trip delay constraints. SLAs for instance are negotiated and crafted in this layer.

Section 1.3 that follows applies to the optical layer as defined in this section. It deals with the wavelength switching and wavelength grooming functions, their distribution across equipment and layers, their interplay, and their impact in terms of transport efficiency and transport failure recovery performance.

1.3   Multi-Tier Optical Layer

An opaque core optical switch, as described in Section 1.1.2, converts optical signals into the electrical domain at the ingress port, switches the electrical signals through an electrical switch matrix, and then converts signals into the optical domain at the egress port. The switch fabric of the OEO switch is typically strictly nonblocking, allowing any interconnection pattern among the ports of the switch (e.g., a Clos switch fabric [76]). This is not always the case though, as some of the switch fabrics for an OEO switch can also be rearrangeably nonblocking (e.g., a Benes switch fabric [35]) or wide-sense nonblocking. The reader is referred to [154] for a comprehensive review of the switch interconnection fabrics.

An OEO switch performs aggregation and grooming functions. As an aggregation device, the OEO switch takes multiple bit-streams and maps them onto wavelengths. As a grooming device, the OEO switch can interchange time slots between different bit-streams. If the OEO switch can switch at the STS-N (Synchronous Transport Signal[6] level N) granularity, we call it an STS-N switch, and if it can switch at the STS-M granularity, we call it an STS-M switch. For example, an STS-M switch can aggregate STS-N traffic onto OC-N wavelengths ($M < N$, N a multiple of M), and can switch STS-M frames between OC-N wavelengths. However, the STS-M switch cannot switch STS-K ($K < M$, K a multiple of M) frames between different STS-N frames.

Figure 1.13 illustrates the aggregation and switching functions of an OEO switch. Switch A aggregates frames from four Input/Output (I/O) interfaces and multiplexes them onto a wavelength

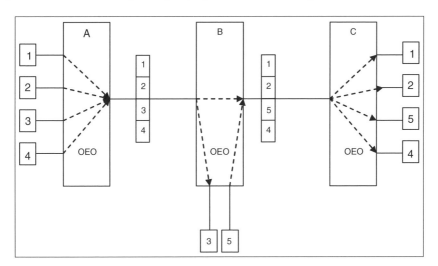

Figure 1.13: Grooming (aggregation and switching) performed by an OEO switch.

[6]STS-1 (Synchronous Transport Signal level 1) is the basic building block of SONET. STS-N signals are created by concatenating multiple STS-1 signals or via a combination of other concatenated signals STS-M with $M < N$ and N a multiple of M. Before transmission, the STS-N signal is converted to an OC-N (Optical Carrier level N) signal. Refer to [4, 90] for additional details on the SONET protocol.

channel. Switch B drops frame 3, and adds frame 4 and multiplexes them onto a wavelength channel to switch C. Switch C demultiplexes the frames onto the I/O interfaces. The granularity of the switches must be at least the frame-rate.

Historically, as the network has evolved so has the granularity of grooming. In the early days of the transport networks, the grooming granularity at the core was 64 kbps (DS0 – Digital Signal 0 is the lowest level of the Plesiochronous Digital Hierarchy (PDH) system [90]). Over time as networks grew and traffic volume increased, the grooming granularity also increased to 1.5 Mbps (DS1) and then to $\approx$ 45 Mbps (DS3/STS-1) to keep the networks scalable and manageable and to improve network performance and cost. As Digital Cross-Connect Systems (DCSs) have been introduced into digital core transport networks over the past 25 years, the rate of the core transmission speed has traditionally been about 20 to 40 times the rate of the *core switching* (cross-connect) rate. DS0 (64 kbps) signals were switched when core transport systems were on the order of DS1 signals (1.5 Mbps); similarly, DS1 signals were switched within DS3 ($\approx$45 Mbps) signals. Most recently, DS3 signals were switched with DCSs when the core transmission speeds were on the order of 1.5 Gbps to 3 Gbps ($\approx$STS-48) [124].

The right granularity for grooming at the core has been a question that has been continuously investigated by engineers. There are advantages and drawbacks for either of the following approaches: switching with fine (e.g., STS-1) granularity (e.g., offering enhanced flexibility to manage all services in the network) or switching with coarse (e.g., STS-N) granularity (e.g., offering increased manageability and scalability and keeping the network complexity under control). Clearly, the drivers behind such a decision will be the expected growth in traffic volume, the traffic composition (emerging applications), the need for scalable and manageable networks, the performance of the network in terms of service provisioning and recovery, the advances in enabling transport network technology, and finally the total network cost [124]. These are factors that need to be considered very carefully before a decision on the grooming granularity is taken.

Apart from the granularity of the network switching elements, another issue of crucial importance is the architecture of the core mesh optical networks in terms of layering in the optical domain. There are two possible architectures for the backbone network: a *flat* (one-tier) architecture or a *layered* or *hierarchical* (multi-tier) architecture.[7] Historically, large networks have always been organized in multi-level hierarchies. It has been a network provider's dream to accommodate all services at all rates with a single box that is scalable, manageable, and low-cost. However, practical considerations such as hardware and software scalability and manageability have led mostly to hierarchical network architectures, taking advantage of the optimization of each layer independently. *All-purpose boxes* may be well suited for enterprise and some metro applications, but typically not for core applications that require specialized, carrier grade products. In a hierarchical architecture, scalability and manageability are achieved by multiplexing traffic flows into larger streams as they traverse from the edge to the core of the network and demultiplexing them as they traverse from the core to the edge. In other words, traffic flows are groomed at a coarser granularity at the network core than at the edges of the network [194, 216, 334].

The sections that follow will define each of these network architectures and will analyze them, trying to identify under what conditions each of them should be used.

1.3.1 One-Tier Network Architecture

The discussion in this section (and in the next section on two-tier network architectures) uses STS-1 and STS-48 as the explicit notations of granularity in order to more easily explain the grooming and multi-tier concepts to the reader. The general argument could be applied for networks with any STS-M and STS-K switches (e.g., $M < K$).

Figure 1.14 shows a sketch of the one-tier network architecture. Ubiquitous STS-1 switches in different offices are connected through physical links. DWDM systems carry the physical connections

---

[7]This layering applies to the optical layer described in Section 1.2.1.

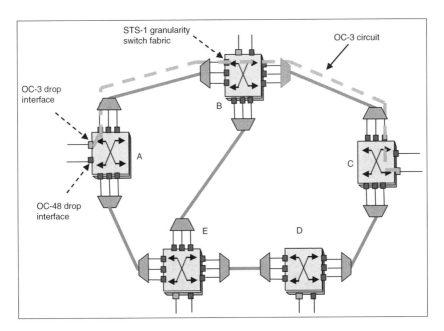

Figure 1.14: One-tier network architecture. (From [189], Figure 3. Reproduced by permission of © 2003 The Optical Society of America.)

from one node to another. In the one-tier (flat) architecture, the core optical switch can switch at the STS-1 granularity. The term *one-tier* comes from the fact that for the STS-1 traffic, the network looks *flat*, i.e., STS-1 traffic in principle is switched end-to-end on the shortest path across the network. However, it can be argued that for rates below STS-1 (e.g., DS3) we need a second tier of (subrate) switches. Nevertheless, we will call it a one-tier architecture, since our focus is the core optical network where traffic rates at or above STS-1 dominate. The STS-1 switch handles all the STS-N services (whose rates are multiples of the STS-1 rate) from the client equipment (such as IP routers, ATM switches, Frame Relay (FR) switches, MSPPs). The STS-1 switch also terminates wavelengths (e.g., OC-48/OC-192) from the DWDM equipment as illustrated in Figure 1.14. In general, an STS-1 switch can switch STS-1 frames from the ingress bit-streams onto egress bit-streams, and allows wavelengths to be managed in increments of STS-1.

An STS-1 switch fabric may be implemented using both space-division and time-division multiplexing. The STS-1 granularity of the switch fabric allows the switch to terminate STS-N traffic with interfaces that can multiplex/demultiplex STS-N traffic onto the switch fabric. A consequence of the STS-1 granularity of the switch is that multiplexing several STS-1 streams from different inputs to a single output could involve the configuration of a large number of cross-point configurations (routing across the switch fabric, and setting appropriate time slot switches). The availability of dense time and space division cross-point chips allows such a switch fabric to scale theoretically to 2000 × 2000 I/O ports.

Figure 1.14 illustrates an example of provisioning an OC-3 circuit from backbone node A to backbone node C. A route is first found across the optical network, with available bandwidth on each link of the route. Then each switch can be configured to set up the OC-3 circuit.

Note that an OC-N circuit between regional PoPs that has to traverse the core optical network is provisioned with multiple legs. The first leg is an OC-N circuit that traverses the first regional network onto the core optical switch at the backbone node. The second leg is an OC-N circuit between

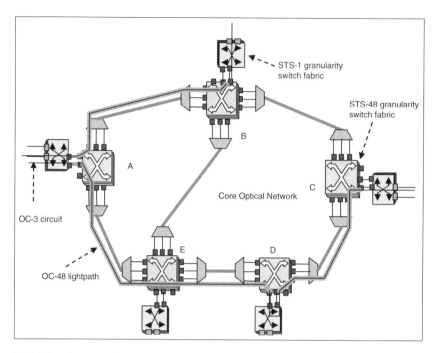

Figure 1.15: Two-tier network architecture. (From [189], Figure 4. Reproduced by permission of © 2003 The Optical Society of America.)

two backbone core switches. The third leg is an OC-N circuit from the backbone node core switch, onto the destination regional PoP. Such provisioning in multiple legs is due to the practical constraint that the regional and core networks are controlled by different subnetwork management systems, and recovery of circuits can be performed by different mechanisms on each individual leg of the circuit.

1.3.2   Two-Tier Network Architecture

This section focuses on the special case of two tiers for the multi-tier network architectures. In the two-tier network architecture, shown in Figure 1.15, the core optical switches switch at the STS-48 granularity (perform *core grooming* at STS-48 rates), and connected to a core optical switch are one or more switches that can switch at the STS-1 (or lower) granularity (perform *edge grooming* at STS-1 rates). The STS-48 switches terminate OC-48 and OC-192 services, and the STS-1 switches terminate STS-N services below STS-48. The STS-48 switches also groom STS-48 frames onto STS-192 frames. OC-48 or OC-192 services between backbone nodes are set up as lightpaths by finding a route in the core network, and configuring the STS-48 switches along the route. We term this the two-tier architecture because there is a core STS-48 switching tier, and there is a second tier that switches STS-1 traffic. It can be argued that a third tier is necessary for switching at granularities lower than STS-1, but we will nevertheless consider it two-tier because STS-1 and above granularities dominate the backbone traffic [88, 229].

We could also have a network architecture where the node architecture is not uniform. For example, all nodes could still have an STS-1 level switch, which is essential for handling sub-STS-48 traffic. However, some nodes, either carrying a large amount of add/drop traffic or occupying strategic switching locations, called hub nodes, could also have an STS-48 level switch. This architecture is considered a *mixed two-tier architecture*, because not all nodes in the architecture are of the same kind: some nodes are nonhub nodes, while other nodes are hub nodes. Nodes could now be connected not only through STS-48 switches, but also through neighboring STS-1 switches.

Figure 1.15 illustrates three OC-48 lightpaths (A–D, A–B and D–C), and two OC-3 circuits (A–B, and A–C). The OC-3 circuit A–B rides on the direct lightpath A–B. The OC-3 circuit A–C rides on lightpath A–D, *hairpins* into the STS-1 switch at node D, and rides onto lightpath D–C to its final destination. In this example, the STS-48 switch terminates all OC-48 and OC-192 services on the drop side.

In general, subrate services between backbone nodes are set up as follows: A set of OC-48 or OC-192 lightpaths between the core switches serves as an *overlay* topology for purposes of routing subrate services. For example, initially, the overlay topology may be identical to the physical topology, with a direct lightpath between all neighbors. If there is a direct lightpath between a pair of backbone nodes, and there is enough capacity on that lightpath, a subrate service between the node-pair can use that lightpath. In this case, the STS-1 switch serves the role of a multiplexer/demultiplexer device (as opposed to a switch). For this reason, subrate services that take the direct lightpath between two backbone nodes may be terminated on the STS-1 multiplexer/demultiplexer. If there is no direct lightpath with enough capacity, then the subrate service has to traverse multiple lightpath hops, and at each intermediate node *hairpin* into the STS-1 switch (get regroomed) and get switched onto the lightpath on the next hop. Intermediate grooming is a natural aspect of routing on an overlay topology [231, 298, 335, 336]. Figure 1.15 illustrates this. If the total subrate traffic between a pair of backbone nodes exceeds a threshold, then a direct lightpath may be set up between the pair of nodes terminating on drop ports that are connected to the STS-1 multiplexer, and services are terminated on the STS-1 multiplexer. For those services that cannot be routed on a direct lightpath, they can be terminated at the STS-1 switch, and routed over multiple lightpaths between the STS-1 switches with hairpinning.

The size of the STS-1 multiplexer at a node depends on the total subrate traffic demand, but the size of the STS-1 switch depends only on the number of backbone nodes. Hairpinning is a natural inefficiency of routing on an overlay topology. Hairpinning does not occur in the one-tier architecture because the core switch can switch at the STS-1 granularity. However, the amount of traffic that hairpins is bounded because, in principle, as soon as hairpinned traffic exceeds a threshold, a direct lightpath can be set up between a pair of nodes. In practice, this means that service routes have to be reoptimized periodically which is not an easy task. It is desirable to have subrate services routed on direct lightpaths, while at the same time ensure that network capacity is utilized efficiently by having lightpaths *well-packed*.

Note that if there are more than one STS-1 switch/multiplexer boxes connected to the STS-48 switch, then the network is blocking for STS-N services. A blocking second tier may be acceptable if, for example, each second-tier STS-1 switch/multiplexer terminates services for different customers, or if the STS-N services are expected to be static, and not dynamically changing. If the network needs to be nonblocking for STS-N services, then there has to be a single STS-1 nonblocking switch that terminates all STS-N services at the second tier. The size of the second-tier switch depends on the sub-OC-48 component of the traffic demand, and plays an important role in the two-tier architecture.

Also, to set up subrate services in the two-tier architecture, there needs to be coordination between the STS-48 switches and the STS-1 switches on the management plane. In contrast, in the one-tier architecture, all services can be provisioned using the management system that controls the STS-1 switches. An OC-N circuit between regional PoPs that has to traverse the core optical network is provisioned with multiple legs, just as in the one-tier architecture, with coordination among multiple management systems.

1.3.3 Network Scalability

Although switch fabrics in one-tier architectures that are designed to switch data with finer granularity can theoretically scale to large port numbers, the control and management of the core network (e.g.,

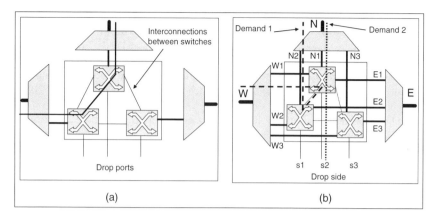

Figure 1.16: Multiple interconnected switches at a single site: (a) Channels from each WDM link are connected to a single switch, (b) Channels from each WDM link are connected to multiple switches.

large port-state databases and a large amount of performance monitoring information), as well as fast mesh recovery at that granularity, will present scalability problems. For example, fast SBPP will be difficult in a one-tier architecture even with bundling, because of the need to perform multiple cross-connects at the end-nodes of the failed link (simulation studies in [16] have shown this). Fast recovery could be achieved utilizing DBPP at finer granularity, however, this incurs the capacity penalty of DBPP. On the other hand, a two-tier architecture using higher-capacity switches is easier to control and manage, and it will be more scalable when recovery is performed at higher granularities.

In addition, as the traffic grows beyond the capacity of the switch over time (something which is more likely to happen with switches of finer granularity), multiple switches will need to be interconnected to yield a larger switch. Otherwise, the network will block some of the connections, or the network capacity will be used inefficiently by routing connections on longer paths. Figure 1.16 shows two possible architectures for multiple interconnected switches at a single site: in Figure 1.16(a) channels from each WDM link are connected to a single switch, whereas in Figure 1.16(b) channels from each WDM link are connected to multiple switches. Both configurations waste ports for interconnecting multiple switches together, as traffic passing through the node can potentially pass through multiple switch ports (interconnect penalty). As the percentage of the traffic that passes through the node is a large portion of the total network traffic,[8] this waste in terms of extra ports used will be significant [254]. In addition, the interconnect penalty is dependent on the traffic forecast accuracy, and it increases when the traffic forecast exhibits considerable uncertainty [189].

Studies show that for arbitrary but uniform set of connection requests, when the switches are interconnected in a mesh topology and 70% of the traffic is pass-through traffic passing through a single switch at the site, approximately 30% of the ports need to be designated as interconnect ports [254]. Furthermore, the design of the interconnection network and the algorithms on how to incrementally add switches and how to route connections in the interconnection network are of crucial importance to the usage of network capacity. Blocking conditions in the interconnection network (the interconnection network may be a blocking network, even though the switches that comprise this network are themselves nonblocking) will lead to stranded or inefficient usage of the network capacity, or a network may not recover from a failure condition, even though there might be enough protection capacity in the network.

---

[8]Typically 70–75% of the traffic that reaches a node will be pass-through traffic. The rest of the traffic will be dropped at the node.

Additional analysis of the one-tier versus the hierarchical architecture was performed in [189]. In the hierarchical architecture the network is scaled by organizing it into layers and these layers are optimized to switch and groom at different rates (STS-1 switching was used for the flat architecture and STS-1 and STS-48 switching for the hierarchical architecture). Studies have shown that there is a crossover point beyond which the layered architecture becomes more cost-effective as the total traffic grows and as the traffic mix evolves towards higher rates [189]. This study assumed uniform traffic pattern (with inaccurate traffic forecast), 30% interconnection ports for the STS-1 switches in the flat network architecture (with nonblocking interconnection), switch sizes in the two-tier network architecture that are not exceeded, and SBPP protection in both cases. Simulations showed that when the proportion of the OC-48 traffic becomes bigger than 50% in the two-tier architecture, that architecture becomes cheaper than the flat topology. In addition, the two-tier becomes cheaper than the flat architecture when the traffic scales beyond the capacity of the STS-1 switch. Similar crossover points were also detected for all unprotected traffic as well.

Furthermore, Capital Expenditure (CapEx) simulations of the flat and hierarchical architectures have shown that the two-tier architecture operating at the STS-48 level in one tier and at the STS-1 level in other, exhibits overall network capital savings on the order of 24−36% compared to the one-tier architecture operating at the STS-1 level [124, 229].

While clearly a number of aspects come into play for a quantitative and fair comparison between flat and hierarchical network architectures, it appears that a layered network exhibits economic efficiency compared to its flat counterpart, as well as improved scalability and network performance.

1.4 The Current State of Optical Networks

The first historical testbeds that were created utilizing optical networking included the Optical Network Technology Consortium (ONTC) [63], Multiwavelength Optical Networking (MONET) [307, 308] and the European Multiwavelength Transport Network (MTWN) [170, 230]. Currently, optical networking has been introduced in both the metro [160] and the long-haul arenas.

Initial testbed experiments [19], and the introduction of network elements such as Reconfigurable Add Drop Multiplexers (ROADMs) [331] have shown the applicability (and cost-effectiveness [267]) of optical networking in the metro space. The typical metro architecture consists of a number of interconnected rings (in a hierarchical fashion) but some mesh network topologies have also appeared. Several testbeds for metro WDM network have been deployed and are described in detail in [226, 301, 303, 333]. For additional architecture and simulation work on designing WDM metro networks the reader is referred to [21, 22, 207, 237, 269, 317], etc.

Long-haul and ultra-long-haul networks initially utilized optical fibers solely as the transmission medium (point-to-point links) and a number of experiments dealt with the enabling technologies and the problem of expanding the reach of these links [223, 302, 311]. With the addition of intelligent and reconfigurable optical cross-connects discussed in Section 1.1, these networks evolved to mesh-based architectures providing enhanced capabilities such as point-and-click provisioning and failure recovery. As analyzed extensively in Section 1.1, these networks can be opaque, transparent or translucent, utilizing various cross-connect switch architectures. Several experiments (and simulations) reported in [239, 266, 283, 309] discuss the design of applicable cross-connects and the viability of these architectures for long-haul and ultra-long-haul systems.

The use of optical networking for long-haul networks has left the laboratory and was successfully commercially implemented in at least three real networks, namely the Dynegy Global Communications nationwide mesh network utilizing Tellium's optical cross-connects (utilizing an opaque design) [66], the AT&T nationwide network utilizing Ciena's optical cross-connects (in an opaque design as well but different than Tellium's) [87, 259], and the Broadwing Communications Services deployment of Corvis' transparent cross-connect. The first two deployments utilized the intelligence of the optical cross-connects to address such issues as point-and-click provisioning, fast failure recovery, and traffic grooming.

The network deployments mentioned above also addressed control and management functionalities of the network, mostly in a proprietary manner. However, there have been significant advances in IETF in developing a control plane for optical networks. An extension of MPLS, namely GMPLS [210], is used for the control plane protocols in optical networks. The three main control functions that are addressed are neighbor discovery, signaling and routing. The reader is referred to [38] for an extensive presentation of MPLS, GMPLS and their traffic engineering extensions. As pointed out in Section 1.1, there are a number of issues that arise during the implementation of control and management functions when completely transparent switch architectures are utilized. A complete analysis of these issues and possible ways to solve these problems in a completely transparent architecture is presented in [108].

Some of the issues that are essential in the successful implementation of mesh optical networks, namely provisioning and failure recovery of connections (lightpaths) and dimensioning of the network, are exactly what the rest of the book is about. The main goal of this book is to present efficient techniques on path routing and single failure recovery[9] in opaque optical networks with arbitrary mesh topologies. These problems are emerging problems in optical networking that have for the large part been dealt with in the research community. This book provides a solid foundation for these problems that can be used by researchers to further their understanding in these issues, as well as by network architects and engineers for the design and implementation of real mesh optical network deployments. The section that follows briefly describes the content of each chapter and the interdependencies between the various chapters in the book.

1.5 Organization of the Book

The advancement of today's optical networks reveals three main trends: arbitrary mesh network topologies, large capacity sessions and the need for reliable services. These trends have motivated us to explore techniques that can provide lightpath provisioning and failure recovery in a fast and efficient manner in mesh optical networks.

This book presents an in-depth treatment of a specific class of optical networks, namely path-protection-oriented mesh optical networks, and focuses on the routing, failure recovery, dimensioning and performance analysis of such networks. Readers who are generally interested in survivability principles are referred to [319] for an extensive analysis of ring-based survivability (with a focus on SONET networks), to [139] for a meticulous account of other mesh-based approaches to survivability, including the p-cycles technique ([139] addresses survivability techniques for optical, ATM, SONET and MPLS networks), and to [305] for a detailed description mainly of the MPLS layer recovery mechanisms.

Furthermore, readers who are interested in optical transmission at the physical layer, or the technologies required for the deployment of intelligent optical networks, are referred to books such as [13, 233], and readers who are interested in optical networking in general are referred to books such as [221, 257, 289].

The rest of the book is organized as follows:

Chapter 2 is devoted mostly to background material on survivability techniques. It motivates the reader with a discussion on failures and the need for survivable networks and presents a survey of the existing optical network fault recovery techniques proposed in the literature. This is by no means a complete listing of all the recovery approaches, but they cover the main techniques that have been proposed over the years. It distinguishes between what are termed protection and restoration

---

[9]The focus of the book is on single failure recovery. Double failures are only considered in Chapter 12, in the context of service availability.

failure recovery techniques and discusses ring, link, path and segment-based survivability approaches. Discussions on multi-layer recovery and integrated protection/restoration approaches in IP-over-WDM networks even though somewhat out of the scope for this book, were added in order to show the direction where survivability research is progressing and to motivate the reader to investigate these areas in more detail.

Chapter 3 is tied directly to Chapter 2 and explores further the classification of fault recovery approaches, focusing on path-based protection techniques for mesh optical networks. The general notion of network components sharing a failure risk is introduced, and Shared Risk Groups (for links, nodes and equipment) are defined and analyzed. Chapter 3 also introduces routing approaches for survivable connections and examines briefly the cases of distributed and centralized routing without discussing the implementation details.

Chapter 4 continues from where Chapter 3 leaves off and describes in detail routing and recovery for the specific case addressed in the remainder of the book, that of failure independent preplanned path-protection for mesh optical networks. The chapter starts with a framework for routing path-protected connections in a mesh network, and then discusses protected connections via the Dedicated Backup Path Protection (DBPP) and Shared Backup Path protection (SBPP) techniques as well as other types of connections such as preemptible, unprotected, etc.

Chapter 5 analyzes the complexity of such routing problems, essentially the complexity of routing working and backup paths in mesh networks, and Chapter 6 introduces, discusses and presents results for various routing algorithms (mostly a variety of heuristic approaches).

Chapter 7 investigates an enhanced algorithm cost model to control trade-offs in provisioning SBPP lightpaths, and Chapter 8 describes three approaches for limiting the number of lightpaths protected by a shared channel for SBPP services in optical mesh networks.

Chapter 9 presents an extension to the computation of SBPP paths using statistical techniques and Chapter 10 investigates lightpath reoptimization and shows how reoptimization offers the network operator the ability to better adapt to the dynamics of the network (demand churn and network changes) that causes the routing to become suboptimal.

Finally, Chapters 11 and 12 address two very timely subjects at this time of writing, namely dimensioning and availability of mesh optical networks. Specifically, Chapter 11 describes analytical approaches to dimension mesh optical networks for backup path protection, and presents techniques that can be used to quickly estimate the network size and failure recovery performance with limited inputs. Chapter 12 ends the book with the modeling and analysis of the service availability mainly of the DBPP and SBPP services, which is a critical tool for the establishment of service level agreements for these services.

The book is organized and demarcated in such a way that readers who may want to just focus on some specific topics can do so without having to read the entire book. For example, readers who are interested generally in survivability can read Chapter 2 and readers who want to gain an insight into path-based protection approaches for mesh optical networks can read Chapters 3 and 4 as well. Readers who want to read through general information on routing algorithms for working and backup paths in mesh networks are referred to Chapters 5 and 6 and readers who are interested in further details on these routing approaches (enhanced cost metrics, limited sharing, routing using probabilistic methods and lightpath reoptimization) are encouraged to read Chapters 7–10. The dimensioning chapter (Chapter 11) and the availability chapter (Chapter 12) can be treated as stand-alone chapters (requiring only some limited background information from the previous chapters) for readers who are interested only in these subjects.

Chapter 2

Recovery in Optical Networks

2.1 Introduction

In currently deployed telecommunications networks, failures occur frequently and sometimes have very serious consequences. As fiber-optic transmission systems are cable-based technologies, they are subject to frequent damage, due mostly to fiber cuts. Apart from construction work (the main cause of failures), fiber cuts may also result because of acts of nature as well as human error [136]. Equipment failures, such as switching node failures, or failures of transmitters, receivers, transponders, amplifiers, etc., are also a source of failure in the network. In general, equipment failures are more catastrophic than fiber cuts, since all the connections passing through the failed network equipment are lost. The source of these failures can again be acts of nature, human error, or hardware/software breakdowns. The reader is referred to Chapter 3 of [139] for an extensive description of causes and impacts of failures in optical networks. As an example, a network link may fail because a back-hoe accidentally ripped a fiber-optic cable, or a network node may fail because a hardware component failed in the switching element of the network node.

The introduction of WDM in commercial networks and the tremendous amount of data that the optical networks can potentially carry have made it imperative for any WDM transport infrastructure to resolve its reliability issues in advance. Fault management of these networks is thus an important problem. Most networks are nowadays required to have the capabilities to quickly detect, isolate and recover from a failure. Research in fault management standards and techniques has been vigorously pursued for many years. This includes design of reliable and robust architectures, failure prevention, detection and identification techniques, conformance testing, verification, fault-tolerant computing and fault recovery techniques. In this book we address only failure recovery. Design of survivable networks, failure detection and isolation and a host of other issues associated with the robust operation of the network are not the focus of this book and therefore they are not dealt with here.

2.2 Failure Recovery

Failure recovery is defined as the process of reestablishing traffic continuity in the event of a failure condition affecting that traffic, by rerouting the signals on diverse facilities after the failure.[1]

---

[1]It is important to emphasize that the term *failure recovery* is used throughout this book as a general term signifying reestablishment of the affected traffic connections after a fault has occurred. This term encompasses both the *protection* and *restoration* terms that appear throughout this book and are defined in Section 2.3.

Path Routing in Mesh Optical Networks   Eric Bouillet, Georgios Ellinas,
Jean-François Labourdette, Ramu Ramamurthy   © 2007 John Wiley & Sons, Ltd

Clearly, failure recovery is a crucial aspect for the successful deployment of today's telecommunications networks. Most users rely on data processing systems and are therefore heavily dependent on their telecommunications networks. Users can range from individual clients to institutions such as hospitals, stock market operations, air-traffic control, banks, sensitive military operations, the retailing industry, schools and government agencies, etc. In many of these institutions, frequent or lengthy periods of service disruption can have severe and even devastating consequences.

A network fault that goes unattended for a long period of time can cause both tangible and intangible losses for the company that provides the service, as well as for its clients. A long outage may cause extensive revenue and asset losses and seriously affect the services provided, thus damaging the credibility and reputation of an organization [318]. A prolonged outage is considered so important nowadays that Service Level Agreements (SLAs) are now the norm between clients and telecommunications carriers. Through these SLAs, carriers now guarantee that the customer will be provided with services with prescribed network availability. Typical availability numbers are of the order of 99.999% (five 9's).[2] This is equivalent to less than 5 minutes of allowable downtime per year. Even though extended outages are particularly harmful, even brief ones can be bothersome, especially when they affect users that are highly reliant on their networks. As an example consider the case of data center connectivity that is used for database replication (e.g., utilized by financial institutions). The customer will not attempt such an operation unless the carrier is able to provide reliable availability levels for this service. Thus the current trend is for more and more networks that are virtually uninterruptible.

We strongly believe that the problem of failure recovery in mesh networks is of crucial importance in the deployment of large-scale core transport optical networks. This book will present solutions for this problem for the general case of mesh optical networks. It is also important for the reader to note that the failure recovery techniques discussed in this book can apply to any circuit-switched mesh network, including IP-over-WDM architectures where the circuits are the MPLS Label Switched Paths (LSPs).

This chapter presents an initial survey of the existing recovery techniques in optical networks and is used to set the stage for the remainder of the book that will discuss in detail one of the recovery approaches presented in this chapter. Readers who are interested in a more extensive and comprehensive analysis of recovery techniques in general are referred to [139].

The rest of the chapter is organized as follows: Section 2.3 presents general fault recovery definitions and classifications, and Section 2.4 describes the protection schemes used in point-to-point SONET architectures. Sections 2.5 to 2.9 describe the ring-based (including SONET Self-healing Rings (SHRs)), path-based, link-based, segment-based and island-based protection techniques used for failure recovery in *mesh* optical networks.[3] Section 2.10 addresses restoration techniques in DCS-based networks, as a foundation for the restoration techniques that could be used in mesh optical networks, and Section 2.11 investigates the case of multi-layer recovery. Section 2.12 discusses recovery triggers and signaling, and concluding remarks are offered in Section 2.13.

2.3   Fault Recovery Classifications

A network is defined as *survivable* if it is capable of failure recovery in the event of a fault occurrence. The degree of survivability is determined by the network's ability to survive single or multiple link or equipment failures. Designing survivable networks is not a topic of discussion in this book as it is a network design problem. The focus of this book is how networks react after a failure condition occurs. In the case where the network is not designed to recover from a failure (e.g., the

---

[2]Some services today may require six 9's (99.9999%) availability as well.

[3]Mesh optical networks are the architectures of interest for the rest of the chapters in the book.

network is one-edge connected[4]), we consider that the affected connections are nonrecoverable and are lost.

Fast and reliable carrier facility/equipment recovery techniques are essential to efficiently protect the network against failures. In addition, for these techniques to be cost-effective, they must also be capacity efficient. The objective of the recovery technique that is employed in a network architecture should be to accurately, rapidly and without great additional cost in terms of redundant capacity, reroute the affected traffic using the redundancy provided in the network, so as to minimize the information lost during the outage. The underlying assumption in survivable networks is that enough redundancy is present in the network in the form of idle carrier facilities or spare capacity in currently used carrier facilities, to ensure recovery from a single link failure. Multiple failures, even though they sometimes occur in the network, are not discussed in this book. The main reason for this is the fact that simultaneous multiple failures occur very infrequently. For a network recovery technique to guarantee against two or more simultaneous failure scenarios in essence means that significantly more redundant capacity will be required to guard against failure conditions that are very unlikely to occur.

In the first half of the 1990s most of the work on fault recovery techniques in optical networks focused on point-to-point systems and SHRs, as natural extensions of the SONET recovery techniques. Protection methodologies used in SONET point-to-point and ring architectures, which were widely utilized and recognized as simple and reliable, were initially adopted to protect against failures in similar WDM point-to-point and ring architectures. A short description of point-to-point and ring SONET protection architectures is included in Sections 2.4 and 2.5 that follow. The reader is referred to [139, 289, 319], for a complete description of protection techniques for point-to-point and ring architectures for both SONET as well as optical networks.

Recovery techniques for mesh networks initially included only the centralized and distributed approaches in networks using Digital Cross-connect Systems (DCSs).[5] It was only in the second half of the 1990s and beyond that recovery techniques in arbitrary mesh optical networks were seriously considered and several research papers were published. This area of research was finally implemented in commercial equipment that was part of real fiber-optic networks starting with Tellium's implementation of Shared Backup Path Protection (SBPP) for their Aurora[TM] optical cross-connect that was used in Dynegy's network implementation [65] and Ciena's implementation of another form of SBPP for their CoreDirector[TM] optical cross-connect that was used in AT&T's network implementation [87].

Different recovery methods have their advantages and disadvantages. For example, a recovery method can be very fast but it can use excessive redundant capacity. Another method can be slower but can use redundant capacity very efficiently. There are several metrics that can be used to evaluate the performance of a recovery technique. Some of them are speed of recovery, capacity efficiency, cost, number of signaling messages exchanged, etc. Clearly, each customer who is using a service has different requirements on the level of protection that it requires for its service. Service providers (carriers) can offer several types of services in their (optical) networks, and each service is associated with a different service level agreement (SLA). Depending on the customers' needs, and how critical their services are (their applications can range from voice and video to critical data transfers), the carriers offer SLAs that bind them to provide services to the customer with a prescribed quality and availability [130]. In optical networks, the services that are offered include among others:

- Guaranteed fast recovery service (50 ms recovery time) using dedicated $(1+1)$ diverse routing (DBPP),

- SBPP protected service (a few hundred ms recovery time),

---

[4]An one-edge connected network is defined as a network where the removal of one link breaks the network into two disconnected parts.

[5]Section 2.10 reviews some of the most important restoration techniques currently used in mesh DCS-based networks.

- Services with multiple diverse paths,

- Unprotected (non-preemptible) services,

- Best effort (preemptible) services using the redundant capacity in the network.

Before we go any further into the description of recovery approaches in Wavelength Division Multiplexed (WDM) optical networks with arbitrary mesh topologies, let us first try to offer some definitions to describe the main survivability concepts that will be brought forward in the remainder of this chapter. The reader should note that there are a number of different recovery definitions and classifications that have appeared in the literature. The survivability definitions presented here are the ones that have prevailed in the research community in recent years. Further classification of recovery techniques (centralized vs distributed route computation, node vs link failure protection, etc.) will be discussed in Chapter 3.

- **Protection** signifies recovery techniques where the backup path (the alternate path that the affected signal takes after a failure condition) and backup channels are *precomputed* prior to the failure occurrence. It is important to note that if the backup path is precomputed but not *preconfigured* prior to the failure occurrence this is also considered a protection technique.[6] For example, recovery techniques used in point-to-point and Self-Healing Ring (SHR) architectures (WDM or SONET-based) that are based on 100% redundancy and physical layer Automatic Protection Switching (APS) to automatically switch the traffic from working to protection facilities are classified as protection techniques.

- **Restoration** signifies recovery techniques where the backup path and backup channel are *not precomputed* prior to the failure occurrence but rather are calculated in real time after a failure has occurred. Switching equipment and spare capacity in conjunction with a rerouting scheme are then used to reroute traffic in the event of a failure.[7] Restoration techniques do not require specified redundant facilities to carry out failure recovery, but they depend on redundant capacity available in the existing carrier facilities. Recovery is provided through route reconfiguration around the failure using the intelligence that resides at a centralized network manager or at individual network switching nodes.

- **Ring-Based Protection** in mesh networks refers to the usage of pre-computed cycles in mesh networks to reroute the signal around a failure similar to the SONET SHR approach [321].

- **Link or Span-Based Protection/Restoration** in mesh networks refers to the rerouting of the failed connection only around the failed link (local protection/restoration provided by the end-nodes of the failed link). In its more general form, when a failure affects more than one working channel, the affected channels recover from failure by using possibly several distinct routes between the end-nodes of the failure.

- **Channel Protection** in mesh networks refers to the case where a spare channel is used on the same link (on a channel on a span along the primary path) for failure recovery (after for example a transmitter failure).[8]

---

[6]Some publications refer to recovery techniques where the backup path is precomputed but not preconfigured prior to the failure occurrence as restoration techniques. In this book precomputation of the backup path is the only requirement used for the classification of the recovery techniques.

[7]Usually these techniques for mesh optical networks are based on reconfigurable optical cross-connects.

[8]This is the more common variety in local span protection, as in the case of APS protection or 4F-BLSR span switching. This is different than the far less commonly used link-based recovery technique described above in which case the recovery path for the local span goes over other spans.

- **Path-Based Protection/Restoration** in mesh networks refers to the end-to-end rerouting of the failed connection. In this case a backup path from the source to the destination of the connection is utilized to bypass the failure.

- **Segment-Based Protection/Restoration** in mesh networks refers to the rerouting of a segment of the failed connection. This is accomplished by using some part of the primary path as part of the backup path. This approach falls between the link and path-based approaches: on one side, if no part of the primary path is used as part of the backup path then we revert to the path-based approach, and on the other side, if the whole of the primary path except the failed link is used as part of the backup path, then we revert to the link-based approach.

- **Dedicated Backup Path Protection (DBPP)** signifies that redundant facilities are dedicated exclusively for the rerouting of a specific connection in the event of the failure on the path of the connection.

- **Shared Back Path Protection (SBPP)** signifies that the redundant facilities are not dedicated for the rerouting of a specific connection, but rather they are shared by a number of different (disjoint) primary paths[9] prior to the failure. After a failure occurs in one of the working paths, that working path then *captures* the previously shared redundant capacity to be used exclusively now for the failure recovery of its signal.

Since protection methods have already precomputed the backup paths and channels, they are clearly faster than restoration schemes and they do not rely on complex computations after the failure occurrence. In general, protection schemes have simple control protocols and allow for service recovery transparent to the users, in contrast to more complex restoration algorithms and protocols for path-based techniques. In the analysis that follows, protection and restoration techniques are described and compared in terms of recovery speed and redundant capacity requirements. This analysis is also a part of Chapter 3, specifically for the case of DBPP versus SBPP techniques.

Figure 2.1 shows a breakdown of the classification of recovery schemes into protection and restoration techniques that can be further classified as dedicated vs shared as well as link vs path vs channel vs segment-based techniques. Similar (or more expanded) tables on recovery technique classifications can be found throughout the literature. Additional classifications of the recovery techniques (preplanned vs real time, centralized vs distributed, failure dependent vs failure independent, etc.) are presented in detail in Chapter 3. The reader should note that, even though it is not reflected in the recovery classification of Figure 2.1, failure recovery is sometimes possible through intra-equipment protection. This is done by switching to redundant equipment such as redundant switch fabric cards, redundant control cards, etc. when a failure is detected.

Figures 2.2 and 2.3 demonstrate examples of link and path-based recovery techniques respectively. It is important to note that even though Figure 2.2 shows that all demands affected by a failure follow a single recovery path around the failure, this, as pointed out in the definition of link-based recovery above, may not always be the case. There are several ways to do link-based recovery and as explained in the definitions above, in its more general form the affected working channels recover from failure by using possibly several distinct routes between the end-nodes of the failure. Finally, Figure 2.4 shows an example of channel protection. In this case the failed lightpath is locally restored by selecting an available channel within the same span. For this technique, all available channels can potentially be used for failure recovery. If no channel is available then other recovery techniques, such as the end-to-end path-based protection can be invoked.

---

[9]The terms *primary paths* and *working paths* are used interexchangeably throughout this book and they signify the original path used to carry the signal information prior to the failure occurrence. The terms *backup path* and *secondary path* are also used interexchangeably throughout this book and they signify the alternate path that is used to carry the signal information after the failure occurrence.

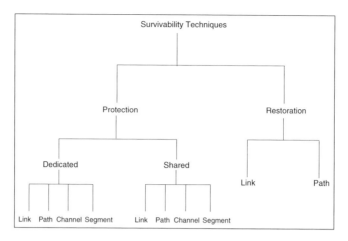

Figure 2.1: Recovery classification.

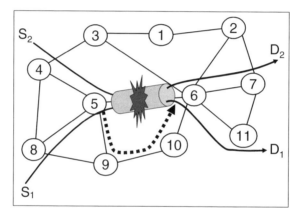

Figure 2.2: Example of one possible link-based protection approach.

Several studies exist on the efficiency of link vs path-based techniques not necessarily in the area of optical networks. For some general discussion on a comparison between these techniques the reader is referred to [163, 164], where the benefits of path over link-based recovery are examined as far as redundant capacity is concerned. Results in that study indeed show that path-based techniques require up to 19% less redundant capacity compared to link-based techniques.[10]

There are many efforts reported in the literature regarding the problems of fault protection and restoration techniques. The rest of the chapter will focus on some of the most important protection and restoration approaches that have been analyzed in the literature in the last decade. Initially, we will describe SONET (Synchronous Optical Network) protection techniques (such as point-to-point, and self-healing ring approaches), as these techniques are the basis for point-to-point and self-healing ring techniques used later in optical networks. Furthermore, we will describe some restoration techniques

---

[10]In networks with a direct link between any two nodes that is always used for a primary path between those two nodes, path-based and link-based techniques become equivalent. Also, because link-based can always be seen as a subset of path-based techniques, path-based techniques will always have better or equal capacity performance.

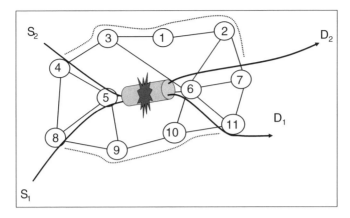

Figure 2.3: Path-based protection.

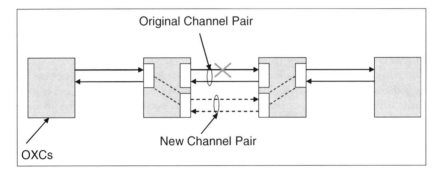

Figure 2.4: Channel protection.

used in networks using DCSs, as these techniques are general enough and they can also be used (with some modifications) in mesh optical networks utilizing optical cross-connects.

2.4 Protection of Point-to-Point Systems

There are three types of Automatic Protection Switching (APS) architectures for point-to-point systems in SONET networks, namely the one-plus-one $(1+1)$, one-for-one $(1:1)$ and M-for-N $(M:N)$ architectures. These protection techniques enable fast recovery (within tens of milliseconds [≤ 50 ms]) with the drawback that they require 100% redundancy (except in the case of $M:N$, $N > M$, where sharing takes place).

2.4.1 $(1+1)$ Protection

In $(1+1)$ protection architectures, there is a backup path for every working path and the $(1+1)$ system provides a diverse route between transmit and receive ends. The traffic is bridged at the transmitter to both paths and one of the two signals is selected at the receiver using a switch. If one of the two paths fails, the switch at the receive end is used to switch to the signal coming from the secondary path, thus recovering from the failure. Since the receiver at the destination detects the failure and switches to the secondary facility independently, signaling for recovery purposes is not required in this architecture. Nevertheless, the SONET APS signaling channel (K_1 and K_2 bytes) is still used to indicate the local switch and the mode of operation [4]. Also, in the $1 + 1$ architecture, the data signal does not revert to the original working path once the failure on that path has been physically repaired. This mode of operation is termed the *nonrevertive mode* [9]. Finally, because of the bridged traffic at the head-end, no low-priority traffic can be transported using the protection facilities (as is the case in the $(1:1)$ architecture discussed below).

2.4.2 (1:1) Protection

One-for-one $(1:1)$ protection architectures are similar to the $(1 + 1)$ architectures described above, with the exception that in this case the traffic is not bridged at the source node but is switched at the working or backup path. Initially, the transmit and receive ends both switch to the primary path. When a failure occurs and is detected by the receive end,[11] and after signaling messages are exchanged between the transmit and receive ends, both ends switch to the secondary path to recover from the failure. Contrary to the $(1 + 1)$ architecture, the $(1:1)$ protection architecture can use its secondary path for carrying extra (preemptible) traffic when the network is in a fault-free state. Upon a failure occurrence, the high priority traffic affected by the failure will switch to the secondary path and the low-priority traffic will be lost. Also, this architecture now operates in a *revertive mode*. When the failure on the working path is physically repaired, the signal will switch back to that path, so that the protection path is again available for failure recovery.

The APS signaling channel is now embedded in the SONET Line Overhead and carried on the secondary path using the K_1/K_2 bytes. These bytes convey the failure message, and trigger as well as coordinate the recovery process. Furthermore, an APS protocol and a number of commands are defined to ensure smooth operation of the protection switching process. These commands, initiated by technicians locally or remotely, can perform various protection switch actions or provision the APS controller [1, 4, 9, 288].

2.4.3 (M:N) Protection

M-for-N (M:N) protection architecture uses the same concepts as the $(1:1)$ approach. However, in (M:N) protection architectures, the protection resources are shared among many working paths by having N secondary paths protecting M working facilities ($N > M$) [149]. If multiple simultaneous failures occur, and there are not enough secondary paths to protect all of them, the working paths with the highest priorities will be protected and the rest will be lost [8, 289]. As in the case of $(1:1)$ protection, this architecture also operates in the revertive mode.

2.5   Ring-Based Protection

Protection of ring architectures has been studied extensively both in SONET as well as WDM networks [115, 321]. In this section we will first briefly explain the recovery techniques used in SONET networks with ring topologies (called self-healing ring [SHR] architectures).[12] The second part of the section describes several ring-based recovery techniques developed for optical mesh networks.

---

[11]Bit Error Rate (BER) thresholds and appropriate time windows are established that initiate the recovery process.

[12]The techniques used later for optical networks with ring topologies, namely UPPR (Unidirectional Path Protected Ring) (or SNCP (Subnetwork Connection Protection)) and SPRING (Shared Protected Ring) (or MS-SPRING (Multiplexed Section Shared Protected Ring)), are similar to the ones used for SONET rings (UPSR (Unidirectional Path Switched Ring) and BLSR (Bidirectional Line Switched Ring) respectively) and will not be addressed here. The reader is encouraged to read [289] for details on these techniques.

2.5.1   Failure Recovery in SONET Networks with Ring Topologies

SONET ring networks consist of SONET multiplexing equipment (Add Drop Multiplexers [ADMs]) interconnected to form closed loops.

The two existing types of SONET SHRs are the Unidirectional Path Switched Ring (UPSR) and the Bidirectional Line Switched Ring (BLSR) architectures. Depending on the kind of protection switching used (path or line, as the names of the two schemes imply) the appropriate SONET layer triggers the protection switching mechanism in the event of a failure [85].

2.5.1.1   Unidirectional Path-Switched Ring

In UPSR architectures, two fibers in opposite directions (clockwise and counterclockwise) are used to connect neighboring nodes on the ring. Each SONET path signal is bridged at the source node to both of the fibers leaving that node, and is transmitted simultaneously to the destination using both directions around the ring [319]. The destination node thus receives signals from both directions and chooses the better of the two (the one with the better signal quality). When a failure occurs, a switch at the receiver is thrown and the signal that is now received at the destination arrives from the other direction around the ring that is not affected by the failure condition. Since the destination node does not need to notify the source node that a failure has occurred, a signaling channel is not required for fault recovery purposes. Clearly this architecture cannot recover from multiple simultaneous failures that affect both directions on the ring, and no low priority (preempted) traffic is allowed to use the secondary (protection) fiber as the signal is permanently bridged at the source node.

2.5.1.2   Bidirectional Line-Switched Ring

BLSR architectures are bidirectional SONET line-based shared protection architectures. Recovery from a failure is now performed at the line layer. BLSR architectures maximize bandwidth utilization and have higher capacity than UPSR architectures for the same traffic patterns. This is the case since BLSR architectures are allowed to reuse the bandwidth and can have additional (preempted) traffic on the protection facilities [321]. There are two architectures for BLSR, the 2-fiber (2F-BLSR) and 4-fiber BLSR (4F-BLSR).

2F-BLSR   Two fibers in opposite directions are used to interconnect neighboring nodes in a 2F-BLSR. Half of the capacity on each of these fibers is reserved for failure recovery [10]. In this case, half of the time slots in each direction are used for the working traffic and the other half are reserved as backup. When a fiber failure occurs, time-slot interchange (TSI) at the SONET ADMs is used to switch the working traffic onto the reserved time slots on the fiber that has not failed (for both directions of the traffic). Therefore, the working traffic is looped onto the fibers going in the opposite direction around the ring away from the failure [139, 289, 319]. (If an ADM on the ring fails, the recovery switching that takes places isolates the failed node completely and sends the traffic away from that node. It is important to note here that traffic that was destined [or originated] for [in] that node cannot be recovered.)

A signaling protocol is required in this architecture. This signaling protocol allows the ring nodes to coordinate the line switch when failure conditions occur. This signaling channel is implemented in the K_1 and K_2 bytes of the SONET Line Overhead, and is exchanged over the protection facilities between SONET Line Terminating Elements (LTEs).

4F-BLSR   The 4F-BLSR architecture uses two pairs of fibers to interconnect the neighboring nodes on the ring (each pair has fibers in opposite directions). Two of these four fibers (one with a clockwise direction and one with a counterclockwise direction) are used to carry the working traffic and the other two fibers are used as backups, to be utilized only in the case of a failure. The 4F-BLSR supports two types of switching during failure recovery, namely *ring* and *span* switching. As in the case of 2F-BLSR, the signaling channel is carried in the K_1/K_2 bytes, on the backup fibers. If a pair of fibers is cut and both the working and backup fibers are disabled, ring switching is used to recover from the failure. When ring switching engages, the failed link is isolated and the traffic uses the backup fibers on the long path around the ring to recover from the failure. If only the fiber that carries the working traffic fails (working fiber), the failure can be recovered by just switching the traffic to the backup fiber on the short path to the destination (i.e., the backup fiber that points in the same direction as the working fiber that failed). This procedure, which is identical to the point-to-point $(1 : 1)$ technique, is called span switching [139, 289, 319].

2.5.2   Ring-Based Failure Recovery in Optical Networks with Mesh Topologies

The protection techniques described previously for SONET ring architectures or variations of these schemes can be applied to mesh network architectures as well, provided that we can find a *ring decomposition* of the mesh architecture, and then use well-established protection switching schemes to restore the traffic whenever failures occur. The three most notable ring-based protection techniques for mesh networks are: (a) ring covers [112, 127, 284] (b) cycle double covers [105, 110, 111, 113, 114] and (c) *p*-cycles [139, 144, 287]. There is also a fourth technique called the generalized loopback technique [117, 118, 204, 214, 213] that is not strictly a ring cover technique, as it does not decompose the network in sets of rings, but we are including it in this section as it uses a *loopback* protection switching mechanism similar to ring-based approaches to reroute the traffic around the failure.

2.5.2.1   Ring Covers of Mesh Networks

The goal of this technique is to find a set of rings that covers all the network links and use these rings to protect the network against failures. As will become apparent below, in this approach some network links (or edges of the graph representation of the network) in the ring cover may be used in more than one ring, which in turn implies that additional redundancy (more than 100%) may be required for the actual implementation of the protection scheme. Minimizing the number of additional protection fibers (i.e., minimizing the redundancy in the network) is the primary focus of such a technique.

Let us now explicitly define what a ring cover of a graph is: given an undirected graph G (with loops and multiple edges allowed), a *ring cover* of G is defined as a set of (not necessarily distinct) cycles, $\mathcal{C}$ such that each edge of G belongs to at least one cycle of $\mathcal{C}$. Thus, a ring cover is a set of cycles (rings) that cover all the edges of a graph. An example of a ring cover of a network is shown in Figure 2.5.

The discussion in this section is limited only to two-edge connected graphs,[13] since graphs with bridges[14] have no ring cover. It was shown in [37] that for bridgeless graphs, a ring cover can always be found. A number of algorithms can be applied in order to find these ring covers. The goal of these algorithms is to find the minimum cost ring cover for any given arbitrary mesh network, so as to minimize the required protection capacity. The minimum cost ring cover in this case is defined as the ring cover with the smallest length.[15]

The simplest algorithm to obtain a ring cover of a network is by finding a set of *fundamental cycles* for the corresponding graph. The fundamental cycles form a basis for the cycle space, and any arbitrary cycle of the graph can be expressed as a linear combination of the fundamental cycles using the ring-sum (exclusive or) operation. Many polynomial time algorithms have been presented in the

---

[13]Graph where two edges have to be removed to disconnect it.

[14]An edge of a connected graph G is called a bridge if it is on no cycle of G.

[15]The *length* of a ring is defined as the number of edges it contains. Given a graph G with ring cover $\mathcal{C}$, the *ring cover length* ($\mathcal{L}(\mathcal{C})$) is the sum of the lengths of all rings in $\mathcal{C}$ [37].

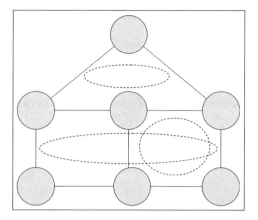

Figure 2.5: Example of a network ring cover.

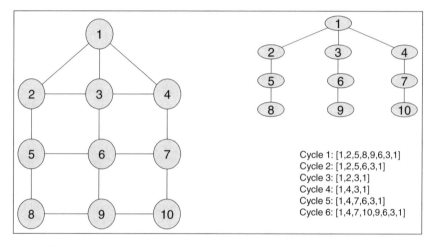

Cycle 1: [1,2,5,8,9,6,3,1]
Cycle 2: [1,2,5,6,3,1]
Cycle 3: [1,2,3,1]
Cycle 4: [1,4,3,1]
Cycle 5: [1,4,7,6,3,1]
Cycle 6: [1,4,7,10,9,6,3,1]

Figure 2.6: Example of a set of fundamental cycles for a mesh network.

literature for finding a set of fundamental cycles for undirected finite graphs [132, 236, 295]. One of the easiest (and fastest) methods of finding the set of fundamental cycles is by using a spanning tree T. This method is as follows: let $G = (V, E)$ be a connected, finite undirected graph with V vertices and E edges and $T \subseteq E$ be a spanning tree, i.e., a maximal subgraph of G without cycles. Each edge in graph G but not in the spanning tree T, $e \in E - T$, when added to the tree, resulting in graph $T \bigcup \{e\}$, yields exactly one cycle. The set of cycles obtained by inserting each of the remaining edges (chords) of G into T will be a fundamental cycle set of G with respect to T. Thus, the fundamental cycle set of G corresponding to T is a set of cycles of G each consisting of an edge (f,g) of $G - T$ together with the unique path (g, ..., f) in T [132, 236]. The number of cycles in the fundamental cycle set is equal to the number of chords added to the spanning tree until all the edges in G are accounted for $(|E(G)| - |V(G)| + 1)$. Figure 2.6 shows an example of a set of fundamental cycles for a 10-node mesh network.

In order to minimize the ring cover length (and thus the redundancy of the network) as much as possible, while using this simple approach, a number of separate spanning trees can be found for each vertex in the graph, and a different ring cover will result from each spanning tree. For a large network, the number of spanning trees and consequently the number of cycles is large enough to prohibit such an exhaustive search [211].

There are a number of algorithms in the literature on how to obtain ring covers of graphs, and more recently a number of efficient algorithms have been proposed to try to minimize the cost of the ring covers in arbitrary mesh topologies [112, 125, 127, 284]. It is left to the reader to further investigate this subject.

Once the ring cover of a graph is found, approaches similar to those used for SONET SHRs can be applied to protect the network against failures. Thus, ring cover techniques provide fast protection (on the order of 50 ms) and are very simple to implement once the rings are identified. If the logical rings are chosen optimally then we can limit the amount of redundant capacity required as well. In the best case, we will require 100% redundancy which is as good a result as SONET ring protection in terms of redundant capacity.

It is important to note here that in this section we addressed only the problem of finding ring covers to be used for fault recovery. This is a simplified approach that does not take into account the problem of jointly routing the demands and finding the rings while trying to minimize the total network cost [313, 314, 315]. Indeed, in that case the ring cover approach will not produce the most efficient results.

2.5.2.2 Cycle Double Covers

Using Cycle Double Covers (CDCs) for the protection of mesh optical networks was first proposed in [114] to alleviate the problems encountered by the ring cover approach (the need for more than 100% redundancy in most cases). The CDC technique provides for a ring cover that requires exactly 100% redundancy, i.e., one protection fiber is required for each working fiber in the network as in the case of SONET SHRs. More specifically, [114] demonstrated that it is always possible to protect against a single link failure in any optical network with arbitrary mesh topology and bidirectional working and protection fiber links, if the protection fibers are decomposed into a family of directed cycles with the following characteristics: all protection fibers are used exactly once and, in any directed cycle, a pair of protection fibers are not used in both directions unless they belong to a bridge.

Cycles with these characteristics can always be found in mesh networks with planar[16] or Eulerian[17] topologies [105]. Any connected planar graph, embedded in a plane, with n vertices ($n \geq 3$) and m edges, has $f = 2 + m - n$ faces [120] where f is denoted as *Euler's number*. The number of faces f includes ($f - 1$) *inner faces* and one *outer face* when the graph is mapped on a plane. The protection cycles are then a set of facial cycles (f boundaries), with each one oriented properly. Thus, for planar graphs the required set of protection cycles can be obtained by embedding that graph on the plane, identifying the faces of the plane graph and traversing those faces in a certain direction (see example in Figure 2.7).

In mesh networks with nonplanar topologies (that are also non-Eulerian), such a set of cycles is also conjectured to exist based on the CDC conjecture which states that for any bridgeless[18] graph, there exists a set of cycles such that each edge is in exactly two of the cycles [166]. Furthermore, a CDC is orientable when it is possible to choose a circular orientation for each cycle of the double cover in such a way that each edge is taken in opposite directions in the two incident cycles of the double cover. An orientable CDC is exactly the goal in this case: a

---

[16]A graph is *planar* if it can be embedded in the plane (or sphere). A graph G is said to be *embedded* in a surface S, when it is drawn on S so that no two edges intersect (have a common point other than a vertex). The resultant embedded graph is called a *plane* graph. The regions defined by the plane graph are called its *faces*, the unbounded region (on a plane) being called the *outer face* [15, 67, 225].

[17]An undirected graph has an Eulerian circuit if and only if it is connected and the number of vertices with odd degree is 0 (or 2).

[18]A bridgeless graph is a graph with no bridges [67].

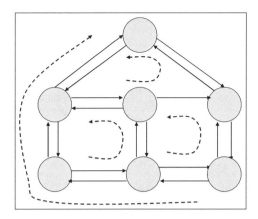

Figure 2.7: Example of a CDC for a planar graph.

cycle decomposition such that each edge appears in exactly two cycles *and* each edge is used in opposite directions in the two cycles (i.e., the set of protection cycles is simply the family of cycles which are face boundaries in the strong orientable embedding that results in this orientable CDC).

There are limited cases where the validity of the CDC conjecture has been proven [61, 165, 297]. While the CDC conjecture has never been proven for arbitrary graphs, it was shown in [166] that a minimum counterexample[19] must be a *snark*. A snark is defined as a cyclically four-edge-connected cubic graph of girth[20] at least five, which has chromatic index four. Furthermore, Celmins in [62] showed that the minimum counterexample to the CDC must be a *strong snark* (a snark G such that for every edge e, $G * e$ (the unique cubic graph homeomorphic to $G - e$) is not edge colorable with three colors). Finally, it was further proven that a minimum counterexample to the CDC conjecture has girth at least seven [131]. Thus, the minimum counterexample to the CDC conjecture has to be a strong snark of girth at least seven. But no snark of girth at least seven is known to exist and it was conjectured in [167] that such snarks do not exist (conjectured that every snark has girth at most 6). Obviously, snarks are graphs with unique topologies and it is not anticipated that telecommunications networks with such topologies will be encountered. So, even if a counterexample to the CDC conjecture does exist, it is highly unlikely that any of the telecommunications networks encountered will be counterexamples. A rule in the design of the network can also be adopted to ensure that this never happens. Work in [105] showed that a ring cover where each network link is covered by exactly one ring can be found for all graphs, provided that an orientable CDC exists for arbitrary graphs. Figure 2.8 shows a CDC consisting of 7 cycles for a 14-node nonplanar graph.

A special case also appears when the undirected graph representation of the original network (which is obtained by replacing both directions of a bidirectional edge with an undirected edge) is Eulerian. Its cycle decomposition S will then provide a family of protection cycles [120]. Figure 2.9 demonstrates a cycle decomposition for a six-node Eulerian graph. Once the protection cycles are identified, failure recovery is possible by using the techniques described above for the case of 4F-BLSR architectures. The reader is referred to [113] for additional details on the implementation of the protection technique once the protection cycles have been identified.

---

[19]If G is a minimum counterexample to the CDC conjecture then G is a bridgeless graph with no double cover which has a minimum number of edges among graphs with these properties.

[20]Girth of a graph G is defined to be the order of the smallest cycle it contains.

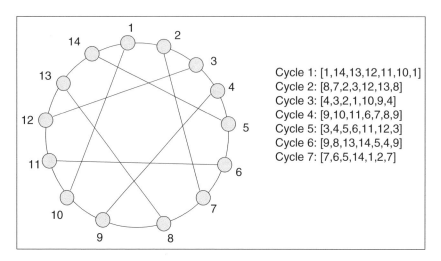

Cycle 1: [1,14,13,12,11,10,1]
Cycle 2: [8,7,2,3,12,13,8]
Cycle 3: [4,3,2,1,10,9,4]
Cycle 4: [9,10,11,6,7,8,9]
Cycle 5: [3,4,5,6,11,12,3]
Cycle 6: [9,8,13,14,5,4,9]
Cycle 7: [7,6,5,14,1,2,7]

Figure 2.8: Example of a CDC for a nonplanar graph.

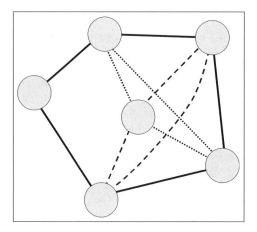

Figure 2.9: Example of a CDC for an Eulerian graph.

2.5.2.3   p-Cycles

The p-cycles approach first introduced in [144] goes one step further than CDCs by allowing the rings (called here p-cycles (preconfigured cycles)) to protect their chords as well.[21] By doing this, less than 100% redundancy is required to protect the mesh network against any link failure, as the straddling spans (chords) have working capacity but require zero units of protection capacity [286, 287]. For example, if the network is Hamiltonian,[22] a single p-cycle is enough to protect all the network links.

The aforementioned p-cycles can protect against a span failure on the cycle or off the cycle. If the span failure is on the cycle then the p-cycle contributes one protection path, and if the span failure

---

[21]A chord is a link spanning the ring.

[22]A graph is Hamiltonian if it contains a Hamiltonian cycle. A Hamiltonian cycle is a cycle that connects all the vertices in a graph, while passing through each one exactly once.

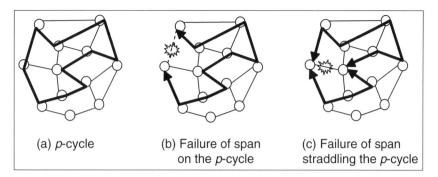

(a) *p*-cycle	(b) Failure of span	(c) Failure of span
	on the *p*-cycle	straddling the *p*-cycle

Figure 2.10: *p*-Cycle example. (From [144], Figure 1. Reproduced by permission of © 1998 The Institute of Electrical and Electronics Engineers.)

is off the cycle (on a chord connecting two nodes on a *p*-cycle), then the *p*-cycle contributes two protection paths. These are the paths around the *p*-cycle that the failed chord belongs to.

There are two types of *p*-cycles, namely *link p-cycles* and *node-encircling p-cycles*. Link *p*-cycles protect all the channels on a link where node-encircling *p*-cycles protect all the connections traversing a node [272]. Figure 2.10 shows an example of a link *p*-cycle and how it recovers from span failures on and off that cycle.

One way to obtain the *p*-cycles in question is to first route the traffic demands and reserve working capacity for these demands and then form *p*-cycles using the spare capacity of the network. *p*-Cycles are chosen such that each working traffic demand is protected by *p*-cycles which have the appropriate capacity. Note that if such a set of *p*-cycles is not found, then a new routing of the working connections is required. Assuming that the network nodes can provide for full wavelength conversion (e.g., opaque network nodes are used), then a mathematical formulation can be used to find the optimal combination of the *p*-cycles. The nodes on each cycle path are pairwise different (cycles are simple) and restricted in length. Since we are assuming that the network nodes are opaque, the length restriction is not imposed for attenuation purposes (as we do not have a transparent lightpath end-to-end) but rather to limit the delay of a connection when it is rerouted on the *p*-cycle after a failure event. Table 2.1 illustrates the notation for the inputs and variables in the Integer Linear Programming (ILP) formulation [272]. The objective function and constraints on the variables are illustrated in Table 2.2. The goal of the ILP is to minimize the spare capacity used in the network by the *p*-cycles. The first constraint gives us the allocation of the protection capacity, while the second ensures that the working capacity is protected. The third constraint in this formulation imposes the capacity restrictions on an edge, and finally the last constraint ensures that the *p*-cycle capacity units are integers. The reader can refer to [139] for a complete treatment of *p*-cycles and the outline of several ILP and Mixed Integer Programming (MIP) formulations for *p*-cycles, examining such cases as maximizing *p*-cycle restorability, jointly optimizing working path routing and *p*-cycle placement, etc.

Various case studies have shown that with *p*-cycle protection, the spare capacity required is similar to that of conventional mesh restoration [144, 287], while still providing fast switching times (comparable to line-switched SONET rings). For example, based on the ILP formulation described above, it was shown that, for *p*-cycles of lengths between 4 and 6 km, the redundant capacity used for the *p*-cycles is about 50% of the capacity compared to the working resources.[23] This scheme has therefore the advantages of both ring-based protection and shared backup path-based protection.

---

[23]This ratio goes to approximately 70% when a wavelength continuity constraint is imposed (i.e., the network nodes are now transparent, not opaque).

Table 2.1: Notation for inputs and variables in the ILP

The following notation describes the inputs and variables of the ILP:

- The network is modeled by graph $G = (V, E)$ with a set of vertices V representing the network nodes and a set of edges E representing the network links.

- P is the set of cycles in the network.

- Each edge j in E contains l_j fibers which in turn contain K wavelengths.

- The cost for a unit of capacity on each edge j is $cost_j$.

- $p_{i,j} = 0$ if edge j does not belong to cycle i and $p_{i,j} = 1$ if edge j belongs to cycle i.

- $x_{i,j} = 0$ if working connection on edge j is not protected by cycle i and $x_{i,j} = 1$ if working connection on edge j is protected by cycle i.

- $c_j = l_j \times |K|$ is the capacity of edge j.

- w_j is the number of working channels on edge j.

- s_j is the number of protection channels used by a p-cycle on edge j.

- n_i is the capacity of p-cycle i.

Table 2.2: Objective and constraints in the ILP

Objective: Minimize the spare capacity used in the network by the p-cycles.

$$\min \sum_{j=1}^{|E|} cost_j s_j$$

Constraints:

$$s_j = \sum_{i=1}^{|P|} p_{i,j} n_i, \forall j \in E$$

$$w_j \leq \sum_{i=1}^{|P|} x_{i,j} n_i, \forall j \in E$$

$$w_j + s_j \leq c_j, \forall j \in E$$

$$n_i \in 0, 1, 2, \ldots, \forall i \in P$$

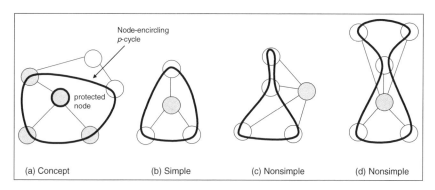

Figure 2.11: Node-encircling p-cycle example. (From [139], Figure 10-49. Reproduced by permission of © 2004 Prentice Hall.)

Theoretical analysis of the efficiency of restorable networks using p-cycles presented in [287] further bolsters the results obtained in various case studies about the spare resources required in these networks. Specifically, [287] proves that the capacity efficiency of a fully preconfigured p-cycle network has the same lower bound on the ratio of spare to working capacity as a span-restorable mesh network, namely $1/(d_{av} - 1)$ where d_{av} is the average span degree of the network nodes.

The analysis presented up to this point was for link p-cycles that deal with the recovery of a link failure. *Node-encircling* p-cycles are the ones responsible for node failure recovery by providing alternate paths among all the nodes adjacent to the node that has failed. Figure 2.11 shows an example of a node-encircling p-cycle. For a p-cycle to be able to protect all network flows passing through a node against a failure of that node, it must contain all the nodes adjacent to the node that has failed, excluding the failed node. Node-encircling p-cycles can be *simple* (a cycle that crosses each node only once) or *nonsimple* (a cycle that crosses some nodes more than once).

Assume that the graph representation of the network in question is biconnected and that the node i (the node to be protected) has as neighbors nodes j, k, m. To generate a node-encircling p-cycle that protects against the failure of node i, we first remove node i and all its incident edges from the graph. Since the original graph was biconnected, removal of a node and its incident edges from it would still result in a connected graph. We then use an algorithm to find the smallest length cycle that includes nodes j, k, m but not node i. These cycles may be simple cycles or nonsimple cycles (i.e., one node-connected with some *bridge* nodes). The reader is referred to [139] for additional information on algorithms for the generation of node-encircling p-cycles.

Some more recent results on p-cycles include flow p-cycles [278] and failure independent path protecting (FIPP) p-cycles [134]. Apart from the span protection capabilities of the original p-cycles, flow p-cycles [278] are now able to provide protection of any flow[24] segment along a path (path segment-protecting p-cycles). Figure 2.12 shows an example of path segments (6−7−2) and (6−7−0) (of paths 10−1 and 9−4 intersecting the cycle) that can be protected by flow p-cycle (0, 2, 3, 5, 6, 8, 0). For example if link 6−7 or node 7 fails, recovery from these failures is possible using the flow-protecting p-cycle. There could be a variety of flows depending on how the path intersects with the flow p-cycle. For example, there could be straddling flows, on-cycle flows, or a combination of straddling and on-cycle flows [278].

Flow p-cycles exhibit redundant capacity requirements that lie between span and path-based approaches, together with the fundamentally fast recovery times provided by the p-cycles technique.

---

[24]Flow is defined as any single contiguous segment of a path [278].

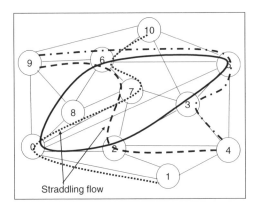

Figure 2.12: Flow *p*-cycles example. (From [278], Figure 2. Reproduced by permission of © 2003 The Institute of Electrical and Electronics Engineers, Inc.)

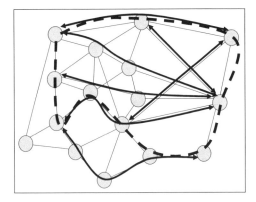

Figure 2.13: FIPP *p*-cycles example. (From [134], Figure 1. Reproduced by permission of © 2005 ICTON.)

For additional details on flow *p*-cycle preselection strategies, and capacity requirements for span and node failures, the reader is referred to [278].

Failure independent path protecting (FIPP) *p*-cycles further extend the *p*-cycle concept to path protection, by including end-to-end *failure independent* protection switching[25] [134]. This approach also protects against span and node failures, but contrary to SBPP techniques, it preselects and preconnects the protection paths prior to the failure. The advantage of such a technique compared to SBPP, is the reduced amount of time it takes to recover from a failure, once the failure is detected, while having a redundant capacity requirement that approaches that of SBPP [134]. A FIPP *p*-cycle will have the following characteristics: (i) it behaves as normal *p*-cycle when the end-nodes of a path are on the cycle, (ii) it protects connections whose working paths are all mutually disjoint or (iii) it protects connections that have diversely routed backup paths on the cycle [134]. This way, when any failure occurs, there is no scenario where two connections will use the *p*-cycle for recovery purposes at the same time. Figure 2.13 illustrates the FIPP *p*-cycle idea by showing a set of connections (shown in solid lines) protected by FIPP *p*-cycle (shown in dashed lines).

---

[25]Contrary to flow *p*-cycles that are failure dependent.

2.5.2.4 Generalized Loopbacks

Generalized loopbacks in the strict sense do not fall under the ring-based approaches; however, as they use a *loopback operation* similar to the APS operation in rings to switch the signal from the working to the redundant capacity, we group them with the ring-based approaches and analyze them here. In generalized loopback the graph representing the network is divided into a primary and secondary subgraph (these are directed and conjugate subgraphs) and use the secondary graph to protect against any failures in the primary graph [213, 214]. Let us assume for simplicity that the network has two fibers per link in opposite directions. The subgraphs are then chosen such that one direction of a link is used in one subgraph and the other direction of the link is used in the other subgraph. If a failure occurs, connections on one wavelength on the primary graph are looped back on the same wavelength around the failure using the secondary graph. The signals that are rerouted propagate through the network by flooding. This approach always guarantees that at least one copy of the affected signal will reach the other side of the failure, as the two graphs are conjugates of each other.

This is a very simple protection technique similar to the CDC approach described earlier. After the primary and secondary graphs are identified, the generalized loopback approach simply reacts in the case of a failure scenario by switching the affected traffic on the secondary graph on one side of the failure and after the signal reaches the other side of the failure it is automatically reinserted back into the primary graph using the same loopback operation. The reader is referred to [117, 118, 213, 214] for additional details on the identification of the two subgraphs in the case of link and node failures and a detailed analysis on the performance of such a technique. Figure 2.14 provides an example of a primary/secondary graph for a sample network and demonstrates how it protects the network against a link failure.

2.6 Path-Based Protection

2.6.1 Dedicated Backup Path Protection (DBPP) in Mesh Networks

In DBPP, the traffic for each connection is sent from the source to the destination node via two disjoint paths (namely the working and backup paths) through the arbitrary mesh network. When a failure occurs, since the traffic is bridged at the source node, no signaling is required to switch it to the backup path. Propagation of the failure to the destination node triggers a protection switch from the primary (working) path to the secondary (backup) path. This operation uses the same

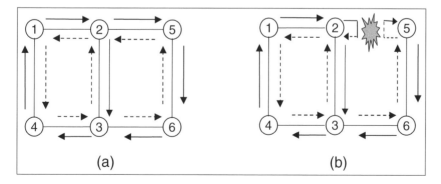

Figure 2.14: Generalized loopback example. (a) Primary (solid) and secondary (dashed) graphs, (b) Protection after a link failure.

principles as the point-to-point $(1 + 1)$ Automatic Protection Switching (APS) in WDM or SONET systems [8, 9, 50, 288, 319, 320] and is the simplest form of path-based protection possible in mesh networks. (This technique is also identical to Unidirectional Path-Switched Ring (UPSR) [319] used in ring SONET networks and Subnetwork Connection Protection (SNCP) [23] used in ring optical networks.) The two paths can be link (fiber)-disjoint, link and node disjoint, as well as shared risk group (SRG)-disjoint [107, 187, 250] depending on the network entities that we are trying to protect. For example, recovery from link failures only requires that the working and protection paths are link-disjoint, whereas recovery from link and node failures requires that the working and backup paths are now link and node disjoint. The term SRG that was mentioned above simply denotes a group of entities, these being fibers in a single cable, cables in the same conduit, colocated nodes, etc. that share a common risk, meaning that they can be affected by the same failure [107, 250, 290] (extensive discussion on SRGs appears in Chapter 3 that follows). Clearly, two entities that are in the same SRG cannot be on the working and its corresponding protection path, as a failure affecting both of them will render the connection unrecoverable. Algorithms to obtain diverse paths with different diversity requirements can be found in a number of publications and research articles. For example, [40] offers a wide variety of techniques for finding disjoint paths with a number of constraints, while [293] is an example of a seminal research article on the same subject. Elaborate discussion on these algorithms is included later in this book (Chapters 6 and 7) and is not the focus of this chapter.

A simple illustration of the DBPP architecture is illustrated in Figure 2.15. Clearly, the advantage of this technique is that it is extremely fast (recovery on the order of a few milliseconds) and extremely simple. It is the preferred method of protecting a connection through a mesh network when recovery speed is of paramount importance. However, it makes inefficient use of capacity, since 100% redundancy is required for protection.

2.6.2   Shared Back Path Protection (SBPP) in Mesh Networks

In SBPP, disjoint working and backup paths are precomputed for each traffic demand as in the case of DBPP [40, 293]. However, contrary to dedicated protection, the redundant capacity on the backup path is not exclusively dedicated to a specific traffic demand but rather it is *shared* among a number of working paths. This in essence means that the signal is not bridged onto the backup path as in the case of DBPP. Rather, the redundant capacity is only *soft-reserved*, and the switching elements along the backup path are not configured prior to the failure [51, 55, 89, 93, 107, 187]. By having multiple working paths sharing the same redundant capacity, we obviously gain in terms of redundant capacity requirements for the entire network. A question then that naturally arises is when can two backup paths share the same redundant capacity? Clearly, this is only possible

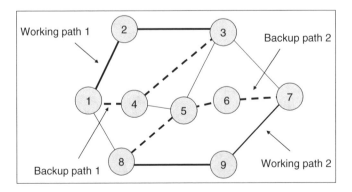

Figure 2.15: DBPP architecture.

when their respective working paths are disjoint [107], as this implies that a failure will not force two working paths to contend for the same reserved redundant capacity. In that case, the level of disjointness (link, node, SRG, etc.) and the recovery requirements (recovery against link failures only, recovery against link and node failures, etc.) will determine the extent of the shareability that is possible in the network. Even though this approach is more capacity efficient than dedicated protection, [51, 55, 68] it is more complex, as it requires the exchange of signaling messages after the failure event in order to configure the switching elements and *seize* the redundant capacity that was soft-reserved. This in turn introduces time delays (time to send signaling messages and time to configure all appropriate switching elements on the backup path) that make such a technique slower than the DBPP approach. An example of the SBPP technique is shown in Figure 2.16. Two working paths, namely WP_1 $(1, 2, 3, 7)$ and working path WP_2 $(1, 8, 9, 7)$ share backup path $(1, 4, 5, 6, 7)$, as they are link and node disjoint. If for example, link $(2 - 3)$ fails, after appropriate failure detection and notification processes, signaling messages will be sent along the backup path to configure the switching elements in order to bridge the traffic from working path WP_1 to the backup path and to allow the traffic to flow along the backup path from the source to the destination.

Similar to SONET (M : N) APS architecture, SBPP will also operate in a revertive mode. This is in general true for all architectures where the backup path is shared among a number of working connections. The main consideration in this mode of operation is to keep the disruption of the traffic to a minimum when the switch back to the original working path is performed. In this case, a bridge-and-roll technique will be used. The connection that has been switched to the backup path, upon the detection of a failure on the working path, will be bridged (at the source node) to the old working path once that path has been physically repaired. When the connection on the original working path has been successfully bridged, the data is rolled to the new path by means of switching at the destination node. The backup path is then torn down and the redundant capacity is again available for failure recovery. A more detailed discussion and analysis of this technique which is the main focus of the book is included in the chapters that follow.

For both DBPP and SBPP, the provisioning of diversely routed end-to-end working and backup paths is performed dynamically during automated provisioning. A different provisioning approach is the protected working capacity envelope (PWCE) concept presented in [133]. This approach is based on the concept that the provisioning and protection mechanisms are now decoupled. PWCE involves utilizing the predefined redundant capacity available in the network and a recovery scheme to create an *envelope* of protected working capacity on every link. When provisioning a service,

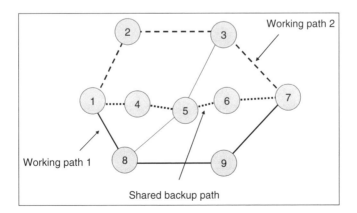

Figure 2.16: SBPP example.

this technique does not provision a backup path for every working connection, but it rather finds the shortest path through the envelope. Any of the span-based protection methods outlined in this chapter, i.e., ring-based protection (e.g., ring covers, CDCs, *p*-cycles, generalized loopbacks) can now be utilized to recover from the failure. The motivation behind this reasoning was to decrease the signaling information exchanged during provisioning and to lessen the maintenance of the databases at each node (including shareability information, etc.) that are required for routing purposes of the protected connections (especially for highly loaded networks with large traffic churn), thus making the network more scalable. An outline of the benefits of such an approach and a comparison of its characteristics with those of SBPP can be found in [133].

2.7 Link/Span-Based Protection

In local span-based (or link-based) protection utilizing OXCs, whenever a failure is detected, the optical nodes closest to the failure attempt to reroute the lightpaths through alternate circuits around the failure (Figure 2.17). In its general form, the affected paths can follow different reroutes between the endpoints of the failure. Some examples of link-based protection architectures are SONET Bidirectional Line-Switched Rings (BLSRs) architectures [319] and mesh network architectures utilizing ring cover, *p*-cycles [139] and CDC [113] protection that were analyzed in the previous sections.

In link-based protection techniques, protection is initiated by the nodes that bookend the failed link. These nodes are the ones that detect the fault (a common detection scheme for all-optical networks could be to detect loss of light) and they are the ones that initiate the protection mechanism by setting their switches appropriately, so as to detour the affected traffic from the failed link onto predetermined reserved protection paths [255]. This can be achieved either by using the actual OXCs to detour the traffic or by using protection switches that are dedicated to the implementation of the protection technique.

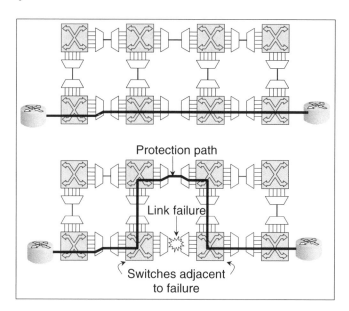

Figure 2.17: Local span-based protection. (From [187], Figure 6. Reproduced by permission of © 2002 Kluwer Academic Publishers.)

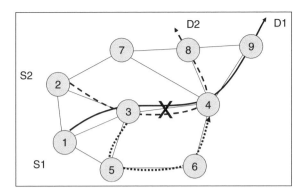

Figure 2.18: Example of span-based protection.

Figure 2.18 illustrates an example of span protection for a mesh network when link (3,4) fails and the failure affects unidirectional traffic from S1 and S2 to D1 and D2 respectively. In this example, the end-nodes of the failed link (nodes 3 and 4) detect the failure, and reconfigure their protection switches so as to detour all traffic that is traversing the failed link. The affected traffic, in this example, is detoured around the fault via a predetermined backup path that uses links (3, 5), (5, 6) and (6, 4). When the detoured traffic from sources S1 and S2 reaches node 4, it continues on its original path towards destinations D1 and D2 respectively, as if a failure has never occurred. In this technique, it is important to note that after the failure is detected, in some cases, initiation of the protection mechanism also involves exchanging protection switching protocol messages between the end-nodes of the failed link prior to the initiation of the protection mechanism.

In this example, for reasons of simplicity, we show that all the affected channels follow the same detour path around the failure. However, as pointed out previously in the definition of the link-based protection approach, in its more general form, when a failure affects more than one working channel, the affected channels recover from failure by using possibly several distinct routes between the end-nodes of the failure.

The reader should also note that in some architectures link (or channel) and path-based recovery mechanisms can coexist, and can be cascaded. For example, for a specific network architecture, channel protection can be the first line of defense. If channel protection fails to recover from the failure (i.e., no channel is available for recovery) then end-to-end path-based protection can be invoked (via path-based recovery triggers that are sent to the end-nodes of the path). Similarly, path-based protection and reprovisioning can also be cascaded. In this case, if path-based protection fails to recover the failure, reprovisioning is invoked and a new working path is found.

2.8 Segment-Based Protection

In the segment shared protection (SPP)[26] technique, a working lightpath is divided into several equal length segments that overlap (Figure 2.19). A disjoint backup path is then found for each working path segment in a similar manner as for regular path-based protection techniques. Shared protection is still possible as before. The main idea in this approach is to limit the size of the working and backup path segments. By doing this we gain in terms of protection speed, as it takes less

---

[26]Also referred to in the literature as short leap shared protection (SLSP) in one of the proposed techniques [157].

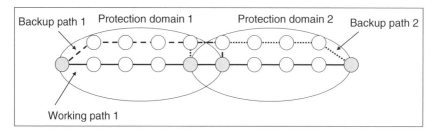

Figure 2.19: SLSP protection scheme.

time for the signaling messages to travel to the end-nodes of the path and fewer switching elements need to be configured. Furthermore, the average time to restore the failure does not vary
with the length of the backup path, but rather it only depends on the size of the segment domains. The computational complexity required to calculate the backup paths is also considerably
reduced, as the working paths are segmented and more resource sharing is now possible since the
SRG constraints are now relaxed (there is no need now to decide on shareability based on the SRG
information of the whole working path). Finally, this approach scales easily with the size of the
network.

2.9   Island-Based Protection

Island-based protection was first proposed in [148] and it is essentially a localized protection scheme
similar to the segment protection approaches presented in [156, 232]. It is, however, a more formal
approach to segment protection, as it formalizes the partitioning scheme.

The basic idea here is to partition the network into smaller subnetworks (*islands*) and to protect
against node or link failures using only the island(s) affected by the failure. As one would expect,
by localizing the protection process the time it takes to recover from a failure is significantly reduced
and the protection times are comparable to link-based protection schemes. Furthermore, by sharing
the protection paths among neighboring islands the capacity utilization is more efficient, approaching
that of the SBPP technique. Thus, island-based protection can allow us to have the best of both the
link and path-based approaches – fast protection times and efficient capacity utilization.

Now let us consider a network node i and the traffic that passes through that node. We can
construct an island I_i that includes node i (called the *island node*), all the incident links to node i,
the nodes adjacent to i and all the links and nodes needed to reroute traffic when node i or any of its
incident links fail. Protection is initiated when a node adjacent to node i detects a failure originating
from any of the links connecting it to node i. In this case, the failure recovery process is independent
of the type of failure (the neighboring nodes do not know whether an incident link or node i itself has
failed). This is one of the main advantages of island-based protection over other protection schemes,
as fault isolation is now not a major part of the protection process.

As a first step, islands have to be identified for all nodes in the network and protection routes have
to be precomputed, together with the allocation of spare capacity. Consider a graph representation of
the network, $G(V, E)$, where V denotes the set of vertices (representing the network nodes) and E
denotes the set of edges (representing the network links). An island I_i is then a subgraph of G with
vertex set V_i and edge set E_i. Island I_i consists of the island vertex i, the set of all edges incident
to the island vertex, the set of all the vertices adjacent to the island vertex, and a set of additional

vertices and edges chosen according to certain island construction criteria, and used to recover the traffic when a failure occurs.

Three types of islands were defined in [147]:

- *Minimal Islands*: A minimal island is defined as an island with the smallest possible set of links required to ensure that the graph stays connected when the island node fails.

- *Shortest-Path Islands*: A shortest path island is constructed by including in the island shortest paths, that do not traverse the island node i, between all the nodes adjacent to i.

- *N-Stage Islands*: An N-stage island is constructed by initially identifying the shortest path island, collapsing that island into a single node and finding the shortest path between the nodes adjacent to the collapsed node. The additional nodes and links that are traversed by these paths are added to the expanded island. This process is repeated N times.

Different types of islands can be constructed based on the desired protection capabilities and characteristics. For example, a minimal island can react to failures very quickly, while a larger island will have a more efficient spare capacity utilization. Figure 2.20 shows three different island constructions for node 22 of a US long-haul network.

Once the islands have been identified, the protection routes have to be pre-computed,[27] and the network's spare capacity has to be allocated for each island. To achieve this, we need to consider

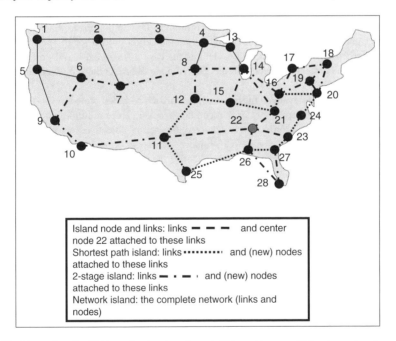

Figure 2.20: Example of a US long-haul network and different types of islands centered on node 22.

---

[27]Protection routes have to follow a number of rules specified by the island protection architecture. These rules are defined in [147].

Table 2.3: Redundant capacity for the island protection technique. (After [147], Table 5.3. Reproduced by permission of © 2003 Columbia University.)

Island type (Size)	Total working capacity	Total protection capacity	Overload protection/working (%)
Shortest hop	2606	2214	84.96
2-Stage	2606	1422	54.57
3-Stage	2606	1302	49.96
Network	2606	1112	42.67

all the relevant failure scenarios within each island. It is important to note that the protection paths always avoid passing through the island node as the process cannot distinguish between node and link failures. Assuming a single failure scenario, it is possible to share the spare capacity on a link among several protection paths using that link. Islands with common links may also share the protection capacity on those links. The reader is referred to [147] for ILP formulations that are used to obtain efficient allocation of the network's spare resources.

Island identification, protection route computation and spare capacity allocation complete the preliminary part of the island-based protection scheme. This is followed by an Island Protection Protocol,[28] which is then used in realtime to protect the network when a failure occurs. Once the failure has been detected, the detecting node informs all the nodes in the island of the failure, so that they can reconfigure their cross-connects accordingly. It is important to note that island-based protection has significantly lower complexity and communication burden for the protection signaling protocols compared to path-based protection. This is the case as in island-based protection only one island takes part in the protection process in the case of node failure and up to two islands in the case of link failures.

Metrics that were used to evaluate this protection approach included protection speed, capacity utilization, number of protection messages exchanged and recovery from multiple failures. Experiments were performed for different routing and capacity optimization objectives, as well as for different types of islands [147]. Table 2.3 shows performance results for capacity utilization for a 24-node mesh network with average node degree 3.33. Clearly, the capacity efficiency of island-based protection improves dramatically as the island size increases, and approaches that of the SBPP approach when the island size expands beyond 2-stages. Figures 2.21 and 2.22 also show that, for a variety of networks, protection speeds are comparable to those of link-based protection-switching systems, and the number of protection messages required is typically at least an order of magnitude less than those required in the SBPP approach.

2.10 Mesh Network Restoration

Restoration approaches specifically for mesh optical networks have not been implemented in real networks, even though some discussion on these techniques has appeared in the literature [256]. The reader should note that even though there are a number of schemes in the literature that are described as restoration techniques, in reality, based on the definition of protection and restoration

---

[28]In [147] this is actually called an Island Restoration Protocol (IRP). For the sake of consistency with the way the terms 'protection' and 'restoration' are used throughout the book, we use the term Island Protection Protocol instead, as according to our definitions this is a protection technique.

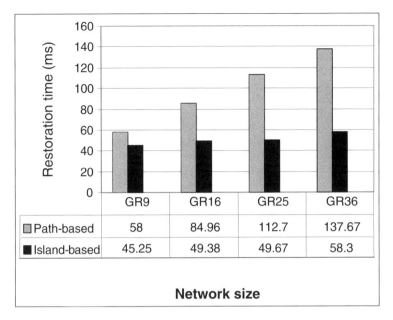

Figure 2.21: Recovery speeds for the island-based protection technique for a variety of mesh networks. (From [147], Table 5.8. Reproduced by permission of © 2003 Columbia University.)

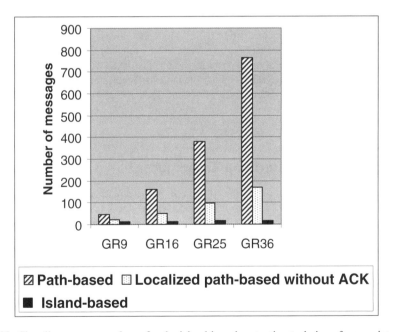

Figure 2.22: Signaling message volume for the island-based protection technique for a variety of mesh networks. (From [147], Table 5.14. Reproduced by permission of © 2003 Columbia University.)

given at the beginning of the chapter, they fall under the protection category. For example, SBPP that was previously described in Section 2.3, is referred to in the literature as a restoration scheme, due to the fact that the cross-connects are reconfigured after the failure has occurred. According to our definition though, since the backup path was precomputed prior to the failure event, and the protection capacity was preallocated, this scheme in truth falls under the protection category, independent of the fact that the actual seizing of the reserved capacity for recovery purposes occurred after the failure event.

Restoration techniques in mesh optical networks are similar to restoration techniques discussed in networks that utilize Digital Cross-connect (DCS) or ATM equipment together with sufficient spare capacity to reroute the traffic when failure conditions occur [163, 164]. In this section we will briefly describe a few of the techniques used in DCS networks as a tutorial on possible restoration techniques for mesh optical networks. These restoration approaches can be very broadly classified as centralized and distributed approaches. In centralized techniques a central management system is responsible for calculating the backup paths and channels after the failure occurrence, whereas in distributed systems, individual switching nodes through the exchange of signaling messages, autonomously discover and configure the backup path. Clearly, centralized systems are less complex and more efficient, as they make decisions based on knowledge of the complete network information, while distributed methods are more scalable and better suited for large networks.

2.10.1   Centralized Restoration Techniques

The central controller utilized in these techniques has access to the complete network informa-tion, including the connectivity, physical topology, available resources, resources utilized, traffic demands, etc. The assumption here is that the information the central controller has is current and accurate. The central controller will then react to a failure notification (through some failure de-tection/identification/notification scheme) and will try to calculate the best possible re-routing path around the failure, based on the information it has about the current state of the network. After the backup path is computed, notifications are sent to all the relevant cross-connects to reconfigure their switching elements so as to accommodate this path. The FAST Automated Restoration (FAS-TAR) and NETSPAR restoration systems among others are examples of centralized restoration sys-tems[29] [64, 82, 265]. AT&T's FASTAR, NETSPAR and MCI's Real Time Restoration (RTR) [176] systems have been in place since the late 1980s protecting the DCS-based networks and restoring DS3s.

2.10.2   Distributed Restoration Techniques

In distributed techniques no central controller is utilized, and a centralized, real-time database image of the network is not required. In addition, no node has a global network description. In these schemes local controllers are utilized at each network node that have only local information, such as information on their switching elements, their connectivity to neighboring nodes and the available and used capacity on the links that are used to connect to their neighbors. In the event of a failure, the local controllers act autonomously in their effort to reroute the affected traffic. Several approaches have been proposed that were mostly a variation of the Self-Healing Network (SHN) technique proposed by Grover in [135] and expanded in [136, 137, 140, 142, 306] for the restoration of high-capacity telecommunications transport networks.

---

[29]FASTAR and NETSPAR restoration systems are in actuality referred to in the literature as a combination of centralized and distributed approaches. FASTAR, for example, uses local controllers to collect and report the failures for fast alarm reporting, and a central controller to calculate the backup path.

In essence this approach was based on the exchange or flooding of messages by the nodes that were directly affected by the link failure. The two network nodes at the ends of the failure designate themselves (arbitrarily) as the Chooser and the Sender and the Sender node is the one that sends on the spare channels of all its incident links a request message. A flooding protocol is used to propagate this message that will finally reach the Chooser node (multiple messages arrive through various paths). One of the paths that is used for the request message to reach the Chooser node is chosen as the restoration path (based on some predefined restoration path criteria such as number of hops, etc.) and a return signature message is sent back on this path to configure the switching elements. Finally, the Sender node sends through the same path the restoration channel ID which is used by the Chooser node to configure its own switching element. The reader is referred to [45, 60, 64, 73, 75, 74, 152, 153, 180, 265, 291, 326] for further discussion on variations of this technique. For discussions on the variations of the SHN approach and the efficiency of each technique the reader is also referred to the Bicknell and Kobrinski studies in [45] and [177] respectively. There, different distributed restoration algorithms were compared using as metrics the restoration time, number of lost working channels restored, utilization of spare capacity, types of failures that can be restored and amount of restoration signaling.

2.11   Multi-Layer Recovery

As described in Section 1.2, the network is in general viewed as a layered architecture. If a failure occurs, a single network layer or a combination of multiple layers can be used for failure recovery in such an architecture [71]. The aim is to provide service protection against a variety of failure conditions while recovering from all failures quickly and with the minimum amount of protection capacity. Failure recovery involving multiple layers can enhance end-to-end recovery of the service by having each layer's recovery scheme supplement each other. It is important to note, however, that multi-layer recovery may not be required or may be difficult to implement because of race conditions and complex escalation strategies and interlayer protocols.

Two competing approaches are being proposed for providing the appropriate recovery mechanisms in these circumstances. In the peer-to-peer approach [39, 243, 244] interweaved optical and higher-layer equipment act in symbiosis under the same control plane. In the overlay approach [243, 244] optical and higher-layer domains are two separate entities with individual control planes, exchanging management services through a standard interface. The peer-to-peer approach relies on a unified bandwidth management protocol to reassign bandwidth away from defective areas in the network and reestablish the interrupted data services. In the overlay approach, each layer independently relies on its own recovery mechanism in a manner that is independent and transparent to one another.

Take as an example the IP-over-WDM networks that are WDM networks that directly support IP (IP/MPLS) and are envisioned as the next generation of networks that will essentially bypass the SONET and ATM layers in the typical network layered hierarchy. Each layer in this architecture provides its own independent recovery scheme. Protection or restoration may be implemented entirely on the IP layer (e.g., IP dynamic routing, Multi Protocol Label switching (MPLS) protection switching) [126, 262], entirely on the optical layer using any of the techniques outlined in the preceding sections, or through some multi-layer coordination efforts between the two layers. The last option outlined usually refers to some escalation strategies between the WDM and IP layers [83, 192] that are discussed later in this section.

A large body of work also exists on creating a common control plane between the IP and WDM layers, which in turn implies that we can now have an integrated (joint) IP/WDM protection/restoration technique [242, 328, 329]. This can be achieved, for example, by extending the MPLS concepts to include wavelength-switched lightpaths. The protection routing approaches can then be broadly divided into two categories, namely sequential and integrated routing. In the sequential case [328], when a request arrives it is initially accommodated on the logical links (existing lightpaths) (both primary and backup paths) at the IP layer. If this is not possible, a new lightpath is created between the edge routers as needed. The integrated (also referred to as hybrid) routing approach uses both logical and physical links (links at both the IP and WDM layers) so as to best accommodate the traffic demand and optimize the usage of the network resources [146, 178].

In the multi-layer networks that are considered in this book, failures are either one of two types: (1) logical, such as a malfunctioning IP-router, or (2) optical, for instance a fiber cut. There are thus four possible scenarios, depending on the origin of the failure, and the layer that provides the recovery [91]:

- Failure and recovery in the optical layer as shown in Figure 2.23(b) based on the original routing shown in Figure 2.23(a). Figure 2.23(a) illustrates the connectivity in the optical layer and the resulting connectivity in the logical layer during normal mode of operation. Figure 2.23(b) is an example of optical failure recovered in the optical layer. The affected lightpath is recovered away from the failure using optical capacity that was reserved for this purpose. The operation is transparent from the logical layer, which remains unchanged.

- Failure and recovery in the logical layer as shown in Figure 2.24(a). This figure illustrates a logical failure (ATM switch or IP/MPLS router failure) recovered in the logical layer. After failure the service is rerouted using the remaining capacity of the logical layer. The operation is transparent to the optical layer.

- Optical failure repaired in the logical layer as shown in Figure 2.24(b). This figure illustrates an optical failure recovered in the logical layer. If the optical layer fails to recover from the optical failure after a certain time lapse, the logical layer can recover the service on a different logical path, using for instance implicit Label Switched Path (LSP) protection in MPLS.

- Logical failure repaired in the optical layer as shown in Figure 2.25. Unlike any of the previous protection schemes, recovering from a logical failure with leverage from the optical layer involves reconfiguration with creation of new connections in the logical and the optical layer. This type of recovery may be necessary if after a logical failure the remaining capacity in the logical layer is insufficient to reroute all the affected services. Additional logical capacity can be created with the provisioning of new lightpaths. This scenario implies a minimum of synergy between the recovery architectures deployed in each layer; the optical layer does not know a priori the logical connectivity of the client and hence cannot take the initiative to recover from a logical failure. Both layers, however, could coordinate their effort to resume

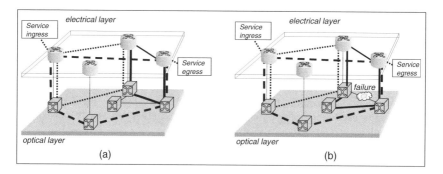

Figure 2.23: (a) Routing before a failure occurs (b) Failure and recovery in the optical layer.

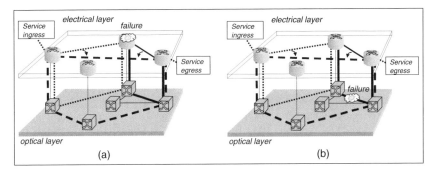

Figure 2.24: (a) Failure and recovery in the logical layer (b) Optical failure recovery in the logical layer.

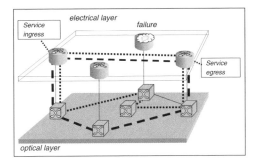

Figure 2.25: Logical failure recovery in the optical layer.

interrupted services, with the optical layer getting directives from the logical layer. In particular, the logical layer could provision spare capacity in the optical domain and reclaim some of it upon failure of one of the routers in order to create new logical connections and balance the load on the surviving routers. Another case is to leverage spare capacity in the optical domain, as well as the optical cross-connects, to route the connections from the primary to a backup router [72, 186].

- A fifth and most realistic situation consists of optical failures repaired simultaneously and independently in both layers. Since this is a combination of scenarios mentioned above, it is not considered here.

Recovery in an IP-centric logical layer is accomplished by Multi-Protocol Label Switching [208]. MPLS enables a hierarchy of LSPs to be defined by prepending a stack of labels or tags to packet headers. Upon an optical or electrical failure occurrence, packets along a given disrupted LSP can be routed to a predefined *recovery LSP* by modifying the label maps of the routers at the endpoints of the original LSP [101]. In a similar manner, recovery of optical failures in the optical layer is also achieved by way of redundancy. Studies indicate that recovery in the optical layer requires substantially more spare capacity, depending on the diligence and the quality of the protection, yet overall the solution is more economical due to lower cost per units of capacity [101]. MPLS offers undeniable potentials for fast recovery. The principal advantage of MPLS is its ability to recover indiscriminately from failures in the logical layer or the optical layer as suggested in Figures 2.24(a) and 2.24(b). However, a single failure may affect thousands of LSPs, and trigger an avalanche of alarms and corrective actions. The resulting amount of signaling can be orders of magnitude higher than in the optical layer, which is able to switch hundreds of LSPs multiplexed into a single wavelength at once. Also in MPLS recovery, primary and backup LSPs must not succumb together to a malfunction in the logical or in the optical layer. In order to satisfy the second condition, the logical layer must explicitly inquire about the risk relationship between the lightpaths that compose its logical connectivity and compute the LSPs, primaries and respective backups, accordingly. Another strategy is to rely on NDP, OSPF and IP self-routing properties to advertise and correct failures in the logical configuration, but then the recovery time is not as attractive in terms of recovery speed as it would be with predefined recovery LSPs.

Experiments described in [139] detail the pitfalls of allowing the faults in the optical layer to escalate to the MPLS layer and utilizing restoration at the MPLS layer to perform failure recovery, simply by reprovisioning the connections that were affected by the failure. Specifically, for one failure scenario shown in [139] that uses mass reprovisioning at the MPLS layer as the recovery mechanism, the restoration levels were at 20% for 22% of the time. Similar results were exhibited with other failure scenarios as well.

To summarize, although scenarios one and three mentioned above address the same problem of recovering from failure in the optical layer, the first, which recovers the failure in the layer where it occurs, is preferable in terms of cost and speed [101, 209]. The same is also true with the second over the fourth scenario. In addition, because the preferred mechanisms are confined within their own layers, that helps simplify the recovery approach, and avoid architectural complexities and interdependence of mixed-layer approaches.

However, notwithstanding the architectural complexities and interdependence of mixed-layer recovery approaches, recovery in different layers could be combined for the best overall result. For instance, fast optical protection architecture for fiber and OXC failures can be supplemented by service-based restoration at the logical layer. In this case, the optical layer can offer bulk recovery of the services while the logical layer can offer finer restoration granularity. If a multi-layer recovery approach is adopted, an escalation strategy has to be provided to coordinate the recovery processes of the different layers. The absence of an escalation strategy can create race conditions between the recovery mechanisms with unpredictable (and potentially catastrophic) results.

The escalation strategies can include either a parallel or a sequential activation of recovery mechanisms. In the parallel approach, recovery mechanisms from different layers are trying to recover the same failure simultaneously, which will result in a very fast recovery time. However, the different recovery mechanisms must be coordinated so as not to obstruct each other or compete for the same redundant resources. In the sequential case, recovery mechanisms from different layers attempt to recover the failure one layer at a time. One sequential approach could be to wait until one layer has failed to recover the services or a fixed time interval has passed before the recovery process is taken over by another layer [11, 33].

Even though multi-layer recovery is clearly an interesting and open area of research, in this book the focus will be only on recovery mechanisms in the optical layer for failures that occur in the same layer. We will not consider recovery mechanisms in higher layers or failures that occur in these layers.

2.12 Recovery Triggers and Signaling Mechanisms

In the preceding sections we described several recovery techniques used in optical networks with mesh topologies. This section touches on some implementation issues for these techniques, namely how is the recovery process triggered and what are the signaling mechanisms that are used during

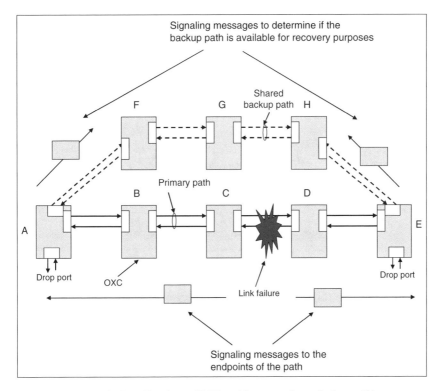

Figure 2.26: Signaling in an SBPP architecture after a fault condition.

fault recovery. To motivate the discussion let us look at the SBPP example in Figure 2.26. A shared backup path is *soft-reserved* for the SBPP primary path, which means that the channels on the backup path may be shared with other backup paths and that the optical cross-connects for the backup path are not set up during provisioning. Instead, the cross-connections are established after a failure occurs and the recovery signaling mechanism initiates.

In this example, the primary path passes through nodes A, B, C, D, E and the backup lightpath spans nodes A, F, G, H, E. We assume that there is a SONET layer on top of the optical layer and that signal failures or signal degradations can be detected using several SONET failure modes such as Loss of Signal (LOS), Loss of Frame (LOF), Alarm Indication Signal-Line (AIS-L), excessive SONET line Bit Error Rate (BER), etc. When link C–D (bidirectional link) on the primary path fails, nodes C and D at the endpoints of the failure detect the failure and recovery triggers are sent to the end-nodes of the primary path. Subsequent end-to-end signaling over the backup path will determine whether the entire backup path is available for recovery. If this is the case, the cross-connects will be set up appropriately, the traffic will be bridged/switched at the endpoints of the working path to the backup path, and the recovery process will be completed. If any part of the backup path is not available (perhaps due to another higher-priority request from another primary path), then the recovery protocol will not complete. In that case the connection will be lost or another mechanism will take over in order to reestablish the connection (e.g., path reprovisioning).

Clearly, defining appropriate triggers and signaling mechanisms, as well as designing a robust recovery protocol, are essential to the efficiency and performance of the recovery process. There can be two types of triggers: triggers from hardware (e.g., Loss of Signal (LOS), Loss of Frame (LOF)

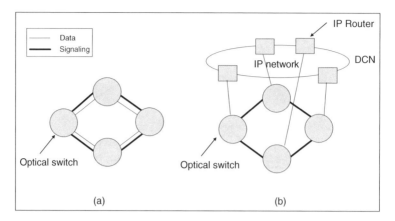

Figure 2.27: In-band vs out-of-band signaling.

for opaque architectures utilizing the SONET over WDM architecture), or control plane triggers. An example of the latter would be to use the control plane to trigger path protection when channel protection fails. Hardware triggers lead to faster recovery of a failure compared to control plane triggers. However, hardware triggers become an issue in transparent architectures where only Loss of Light (LOL) can be detected and where failures propagate to the endpoints of the path (e.g., the failure can be detected at the client's interface and an out-of-band control channel between the transparent switches and the client drop nodes can be used to trigger the recovery process) [109].

Possibilities of signaling mechanisms for failure recovery include in-band bit-oriented approaches (like BLSR), in-band approaches where the signaling messages ride on a protocol like RSVP (or LDP, or Private Network-to-Network Interface (PNNI)) or out-of-band schemes (like LMP) [38]. In the case of in-band signaling the signaling messages are piggybacked on normal data messages, whereas in the case of out-of-band signaling the signaling messages are separated from, and are independent, of the normal data messages. Figure 2.27 demonstrates these two options. In the out-of-band case, a separate channel can be used for signaling and this channel can also be carried in the transport network, or in a completely separate Data Communication Network (DCN). In general, in-band signaling will provide faster recovery than out-band-signaling where latency can become an issue. Also, in-band signaling can significantly reduce the cost of the network since it utilizes exactly the same infrastructure to carry the data and the signaling information. In the case of out-of-band signaling, no bandwidth is taken from the data channels to be used for signaling purposes, thus this scheme allows for the transport of more data at higher speeds. As this approach is completely independent of the data traffic, signaling is also possible at any time. In-band signaling is more natural for opaque architectures but out-of-band signaling will be required to support recovery in transparent optical networks, which will require new out-of-band signaling channels between transparent switches and between transparent switches and client equipment. Note that the latter will also require vendor cooperation and new standards definition [109].

2.13   Conclusion

Network recovery is crucial for the deployment of successful and reliable networks especially due to society's increasing dependence on telecommunications. Recovery of mesh optical networks in particular has received a lot of attention in the last decade mainly due to the widespread use of high-capacity fibers as the means of transporting information and the vulnerability of fiber-optic cables and switching equipment. This chapter initially presented a number of existing SONET point-to-point and self-healing ring protection techniques that are the foundation for some of the mesh optical network protection approaches. It then provided some important classifications of mesh network recovery techniques including protection vs restoration, link vs path vs segment-based techniques, dedicated vs shared, etc. and described the most important of these techniques.

For the link-based protection techniques the focus was on ring-based approaches, and from those we concentrated on ring covers, cycle double covers, p-cycles and generalized loopback. Link-based approaches in general are much faster than path-based approaches but suffer in terms of redundant capacity. Ring covers require more than 100% redundancy, and generalized loopbacks and cycle double covers require exactly 100% redundancy. p-Cycles, however, require less than 100% redundancy and in some cases approach numbers that we usually encounter in path-based shared recovery approaches.

For the path-based protection approaches the focus was on DBPP and SBPP. DBPP techniques are faster and simpler than SBPP schemes, which require more complex signaling to configure the switching elements after a failure event. SBPP approaches, however, are much more economical in terms of redundant capacity required. Segment protection techniques fall between link and path-based approaches. Island-based protection in particular is a recovery technique that requires redundant capacity similar to that of SBPP schemes, exhibits protection speeds similar to link-based protection approaches while generating a considerably smaller number of signaling messages compared to the SBPP approach.

The remainder of the chapter included a discussion on restoration techniques and it distinguished between centralized and distributed restoration approaches. A small section on restoration techniques for mesh networks utilizing DCSs was included as an indication on the types of techniques that could also be used for restoration in mesh optical networks. The chapter also presented a discussion on some of the techniques that can be utilized in order to tackle the multi-layer recovery problem. The chapter ended with some thoughts on recovery process implementation, focusing primarily on recovery triggers and recovery signaling approaches.

This chapter provides a general background on recovery techniques and sets the stage for the rest of the chapters that follow in the book. In the past few years, intelligence in transport switches have allowed service providers to support the same fast failure recovery in mesh optical networks previously available in ring networks while achieving better capacity efficiency and resulting in lower capital cost for mesh networks. Since a number of networks are currently evolving to mesh architectures, as detailed in Chapter 1, the focus in the remainder of the book will be on path routing and failure recovery in exactly this type of architecture. Specifically, we will investigate in detail path routing and failure recovery in mesh optical networks that use the DBPP and SBPP techniques as described in Section 2.6. While the technology the book describes is that of optical networks, most of the concepts and algorithms apply to ATM (Asynchronous Transfer Mode) and IP/MPLS (Internet Protocol/Multi-Protocol Label Switching) networks as well.

Chapter 3

Mesh Routing and Recovery Framework

3.1 Introduction

This chapter continues from where Chapter 2 left off and further classifies the existing protection and restoration types. Even though several protection and restoration concepts are analyzed, special attention is given to protection approaches in networks with shared risk groups (SRGs). For each type of protection and restoration, its distinctive characteristics such as trade-offs of cost versus efficiency, speed of recovery, and other features that set it apart from other types are briefly explained, without discussing the implementation details.

Before entering into the details of the recovery taxonomy, Table 3.1 presents six possible recovery approaches (this is the extension of a similar table presented in [102]), the details of some of which are presented later in this chapter and in subsequent chapters. The table enumerates the three components managed during a failure recovery process. The components are the alternate route around a failure, the channels used along that route, and the embedding of the route into the optical switches. Each category indicates the dependence of each component on the origin of the failure. Components that do not depend on the failure may be assigned before the failure occurs. For components that are assigned after the failure occurs, the table distinguishes between scenarios with precomputed routes but without preassigned channels[1] (categories 3 [29, 89] and 5), scenarios with precomputed routes and preassigned channels (category 4), and scenarios where components are determined and assigned after the failure (category 6)[2] [93]. Categories 4 and 5 depend on the ability of the optical network to perform rapid fault isolation and select the precomputed components from a lookup table or map, based on the location of the fault. Finally, category 7 refers to the SBPP technique with preplanned shared backup path but without reserved bandwidth. In this case the backup paths are precomputed, but there is no reservation of any kind during the provisioning phase. Reservation is implicitly done by network planning, and is not explicit in this architecture.

---

[1]The channels are not precomputed (the routes are) but are assigned during the recovery event from a shared pool. The routing is such that there are enough channels available to recover the paths for any single failure.

[2]The terms 'reprovisioning' (in Table 3.1) and 'path restoration' are technically the same.

Path Routing in Mesh Optical Networks   Eric Bouillet, Georgios Ellinas,
Jean-François Labourdette, Ramu Ramamurthy   © 2007 John Wiley & Sons, Ltd

Table 3.1: Different path-oriented failure recovery categories and their dependence on failure origins. (From [107], Table 1. Reproduced by permission of © 2003 The International Society for Optical Engineering.)

Category	Failure recovery route		Channel assignment on failure recovery route		Cross-connect on failure recovery route	Failure specific?
	Computed	Assigned	Computed	Assigned		
DBPP (Cat.1)	Before	Before	Before	Before	Before	No
SBPP with preassigned channels (Cat. 2)	Before	Before	Before	Before	After	No
SBPP with reserved not preassigned channels (Cat. 3)	Before	Before	After	After	After	Yes (channel only)
SBPP with preplanned maps (routes and channels) (Cat. 4)	Before	After	Before	After	After	Yes (route and channel)
SBPP with preplanned maps (routes only) (Cat. 5)	Before	After	After	After	After	Yes (routes only)
Reprovisioning (Cat. 6)	After	After	After	After	After	Yes (route and channel)
SBPP with preassigned backup paths and no reservation (Cat. 7)	Before	After	After	After	After	No

In the remainder of the book, the cases of precomputed recovery paths and preassigned channels are considered in detail (categories 1 and 2), where the recovery path is the same for all possible failures (as opposed to architectures where a different precomputed path may exist for different failures).

The rest of Chapter 3 is structured as follows. Section 3.2 deals with a more detailed classification of current recovery techniques in mesh networks, while Section 3.3 looks in some detail at the problem of protection in networks with shared risk groups. Section 3.4 discusses centralized versus distributed routing used for protection, and concluding remarks follow in Section 3.5.

3.2 Mesh Protection and Recovery Techniques

As discussed in detail in Chapter 2, starting from the highest level of this hierarchy, recovery against failures can be classified into two main categories, (1) link-based recovery, and (2) path-based recovery. Current recovery schemes in mesh optical networks can also be classified by two more major characteristics: (a) calculation of the recovery route prior to occurrence of a fault (precomputed) vs computation of the recovery route after the occurrence of a fault, and (b) computation of the recovery route prior to occurrence of a fault in a centralized manner, together with centralized recovery protocol implementation after the occurrence of the fault vs computation of the recovery route computation in a distributed manner after the occurrence of a fault and distributed recovery protocol implementation. The sections that follow provide details on each of these recovery techniques and discuss the advantages and disadvantages of each classification.

3.2.1 Link-Based Protection

Chapter 2 described link (or span)-based protection as a recovery method that reroutes the affected channels through alternate circuits around the failure, utilizing the nodes that bookend the failed link. This protection scheme, in general, is faster and has higher availability (compared to a path-based approach), since most of the time only the disabled portion of the path is bypassed, and only the two nodes adjacent to the failed link are involved in the fault recovery process. Even though in its general form link-based recovery implies that the affected working channels recover from failure by using possibly several distinct routes between the end-nodes of the failures, in some cases all of the affected traffic may be detoured using the same backup path. This makes the protection protocol simple to implement, as it performs bulk switching of all traffic demands traversing the failed link, irrespective of their individual source and destination nodes.

On the downside, link-based protection requires, in general, more redundant capacity than path-based schemes [163, 164, 255], and it is very difficult for the network to recover from node failures using this approach. However, there have been as of late span-based recovery techniques, such as the *p*-cycles approach, that have demonstrated redundant capacity requirements close to those required for SBPP [139].

Furthermore, the alternate routes differ for each failure and are difficult to anticipate. The preferred approach to address this problem is to simulate all possible failures and create directive maps stored in the optical switches that assign a precomputed switch configuration for each failure scenario (Categories 4 and 5 in Table 3.1). If a centralized approach is used to achieve this, the generated maps may become very large and this technique will not scale well with the size of the network. The approach also entails lengthy computations whenever a new lightpath is provisioned since it must account for every failure scenario in order to populate the map. For this reason, spare capacity is usually not reserved ahead of time. Instead, the recovery routes are computed on the fly upon a failure event (Category 6 in Table 3.1). This becomes an issue if the failure disrupts many parallel optical channels and sets off a cascade of recovery procedures at the optical switches adjacent to the failure. It is important to note, however, that if a distributed preplanning for span recovery is performed, by having each node store only the failure scenarios and recovery actions that are associated with it (local information within its vicinity only), then this approach potentially scales better than the path-based techniques. In either case, this protection scheme relies on the ability of the network to isolate the failure.

3.2.2 Path-Based Protection

In end-to-end path-based protection, the ingress and egress nodes of the failed optical connection attempt to recover the signal on a predefined backup path, which is SRG-disjoint, or diverse, from the primary path [187, 255, 256]. Path diversity guarantees that primary and backup paths will not simultaneously succumb to a single failure. For example, a backup path that is link-disjoint from its corresponding working path, protects only against link failures. If the restriction is tightened so that these two paths do not share any node, then these paths are said to be node-disjoint and a backup path that is node-disjoint from its corresponding working path protects against both link and node failures.

Unlike local span-based protection, secondary routes are provisioned with the primaries and thus the recovery process does not involve further real-time path computations. Another advantage of path-based protection is that recovery processing is distributed among ingress and egress OXC nodes of all the lightpaths involved in the failure, compared to span-based protection where a comparable amount of processing is executed by a smaller set of OXC nodes adjacent to the failure. In the chapters that follow, only the cases where the backup path is failure-independent and is thus the same for all types of failures are considered. By way of this restriction, the backup paths may be computed and assigned before a failure occurrence (categories 1, 2, and 3 in Table 3.1). There are two subtypes of path-based protection as outlined in Chapter 2, namely DBPP and SBPP.

Contrary to span-based protection, path-based protection has the ability to recover from any single fault, be it a link, node or SRG failure, utilizing the same protection technique. SBPP also enables the minimization of capacity utilization by allowing working paths to share redundant network resources. Furthermore, in path-based protection, the detection of the signal degradation or signal failure takes place at the add-drop ports of the path's egress switches, where access to the overhead bytes is a byproduct of the ports' design (for the case of opaque architectures with opaque switches as described in Chapter 1).

The main disadvantages of path-based protection (especially in the case of SBPP) are (a) it requires a protection signaling protocol, as notification of both end-nodes of the working path is required before the protection mechanism is initiated, and (b) the protection mechanism is performed for each single path that is affected by the failure (no bulk switching is now possible compared to link-based switching where bulk switching after a failure is one of the possibilities). If there is a large number of working paths that traverse a link or node that has failed, this means that a large number of protection instances have to be initiated. This in turn will result in a large number of protection messages being exchanged which will affect the overall time it takes to recover the failed connections. Note that the time it takes to process the protection messages at the optical cross-connects is the main cause of delay in these kinds of networks [16] (analysis the performance of SBPP optical networks in terms of recovery time is presented in Sections 4.5 and 11.8).

3.2.2.1   Precomputed Protection Routing vs Real-Time Restoration Routing

The paths that are used for recovering from the failure can be calculated prior to the failure (precomputed protection case) or after the failure has occurred (real-time restoration case). The discussion that follows analyzes briefly both precomputed protection and real-time restoration routing and identifies the advantages and disadvantages of each technique.

Precomputed Protection Routing: In the precomputed protection case, when a new traffic demand needs to be accommodated in the network, SRG-diverse working and backup paths are computed, and redundant capacity along the backup path is reserved. The precomputed backup path is then stored at the nodes along the backup path (e.g. in the form of routing tables). Protection capacity sharing can occur on the backup paths as long as their corresponding primary paths are SRG-disjoint. This approach corresponds to the SBPP technique. After a failure event, these nodes (based on their stored information) can execute the necessary cross-connects to reroute the traffic onto the appropriate backup path.   If the requirement in the network is for protected services, this means that when a new connection enters the networks, if a backup path cannot be found with enough redundant capacity, then the connection will be blocked. The main advantage of this technique is that the time required for route computation and routing information dissemination is spent prior to the failure event, which in turn means that recovery from a failure will take place much faster after a failure event, compared to the case where the recovery route has to be calculated after the failure event has taken place.

The failure recovery time depends on (i) the time it takes to detect the failure, (ii) the time it takes to calculate the backup path and disseminate the necessary information to the relevant network nodes, (iii) the time it takes for protection signaling messages to be processed at the network nodes (including queueing delays due to multiple messages arriving at a node simultaneously), (iv) the time it takes for the protection signaling messages to propagate in the network, and (v) the time it takes for the nodes to execute the necessary cross-connects. Clearly, by precomputing the backup path, some of the processes enumerated above are completed prior to the failure event, which in turn means that a failure can be recovered much faster compared to the restoration case. A drawback of such an approach is that only protection against a single link or node failure is guaranteed. When multiple failures occur, even if the network remains connected after the multiple failure events, there is a possibility that this approach will not to be able to recover the failure. If, for example, failures occur on two working paths that share the same precomputed backup path, both of them cannot be recovered at the same time, i.e., only one can be accommodated. Another example would be to have simultaneous failures on both the working and its corresponding backup path.

Real-Time Restoration Routing: In the real-time restoration routing case, the restoration path is computed dynamically after the failure occurs, based on the availability of restoration (redundant) capacity at the time of the failure event. As discussed in Chapter 2, in link-based restoration techniques, the nodes that bookend the failure use a flooding mechanism to disseminate restoration signaling messages to each other, via a series of intermediate nodes. This procedure is used to identify the intermediate nodes that will be part of the restoration route. Obviously, this is a time-consuming process, as significant time elapses before the end-nodes can converge to an appropriate restoration route. Parameters that are taken into consideration when calculating the best possible restoration route include, but are not limited to, total capacity utilization on each link, redundant capacity utilization on each link, physical length of the restoration path, number of hops, etc.

If a path-based restoration technique is utilized, the end-nodes of the working path, rather than the end-nodes of the failed link, will now compute the appropriate restoration path. As in the link-based restoration case, signaling messages have to be exchanged between the two nodes, before converging to a restoration path. In both cases, after the restoration path is computed in real time, signaling messages are sent to the intermediate nodes along the restoration path, instructing them to execute the appropriate cross-connects.

Clearly, the main disadvantage of the restoration approach is its slow recovery time compared to the precomputed case. However, this approach is able to accommodate multiple simultaneous link and node failures,[3] as the restoration paths are computed in real time and can use the available restoration resources to recover from multiple failures. This is in contrast to the case of precomputed backup paths where the network is trying to protect against multiple failures by using predetermined backup paths that may not be available after the multiple failure events.

3.2.3 Segment-Based Protection

Section 2.8 described a segment-based protection scheme, that falls between span and path-based protection. The segment protection technique is not limited to the case of shared protection but can be used with dedicated protection as well. If shared segment protection is utilized, it now allows overlapping working (and their corresponding backup) segments and considers the sharing of resources when the segments are determined. The working and their corresponding backup segments have the same end-nodes but they are now node-disjoint. Figure 3.1 illustrates an example of a technique called PROMISE (PROtection using Multiple SEgments) that represents a shared segment protection approach [324, 325]. In this technique the working (active) segments for any given working path follow two constraints: (a) every link along the working path belongs to at least one and at most two working segments, and (b) a working segment cannot be a proper subset of any other working segment. Additionally, each link that belongs to two overlapping active segments is protected only by the backup segment for the second working segment.

The advantage of this approach is that resources are now shared not only among backup segments for different working paths, but also among backup segments for the same working path. Figure 3.1(a)

---

[3]Assuming that the network remains connected after multiple failure events, and that redundant capacity is available in the network for the recovery of multiple failure events.

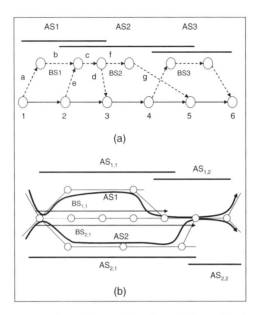

Figure 3.1: Shared segment protection. (From [325], Figure 2. Reproduced by permission of © 2003 The International Society for Optical Engineering.)

demonstrates a case where two backup segments for the same working path (spanning nodes 1 to 5), backup segments BS1 and BS2, share backup bandwidth on link c, and Figure 3.1(b) shows an example of two backup segments ($BS_{1,1}$ and $BS_{2,1}$) that belong to two different working paths and share backup bandwidth as their corresponding *working segments* are link-disjoint. It is important to note here that for the latter two backup segments their corresponding working paths are *not* link-disjoint. Thus, this approach provides additional flexibility compared to the SBPP technique. The flexibility of the shared segment protection schemes also allows for the successful protection of topologies that cannot be protected with either link or path-based approaches (due to the *trap* problem analyzed in Chapter 6).

For efficient algorithms for finding working paths, partitioning them into working segments and determining their corresponding backup segments, the reader is referred to [325]. An ILP formulation is also included in [325] that determines the optimal set of protecting backup segments when the working paths are known. Performance results in [325] have shown that with the proposed heuristics, PROMISE can achieve better results than link-based protection approaches, and no worse results (in terms of bandwidth efficiency) than path-based protection techniques.

Additional work on segment protection is also included in [158] and [159]. In [158], a heuristic algorithm is proposed, called Cascaded Diverse Routing (CDR), where for each source–destination pair, candidate switching/merging node pairs are predefined, and a diverse routing algorithm is used between each switching/merging node pair in order to find the corresponding working and backup segment pair. In [159], another heuristic algorithm called Optimal Protection Domain Allocation (OPDA) is analyzed, where a graph transformation algorithm is invoked to enumerate, select and align simple cycles in the graph as candidate cycles for the working path. A comparison of the PROMISE, CDR and OPDA techniques is presented in [294], where it is shown that the CDR approach yields the best results in terms of the total cost of working and backup path segments compared to the optimal cost achieved by solving the ILP formulation.[4] Finally, [147] describes another form of segment-based

---

[4]The ILP formulation for segment shared protection is also presented in [294].

protection, the *island-based* protection approach. The performance of this technique was discussed in detail in Section 2.9.

3.3 Concept of Shared Risk Groups

In this book, as mentioned previously, and with a few exceptions, only the case of precomputed backup paths independent of the failure (Categories 1 and 2 in Table 3.1) are considered [93, 99, 107, 179, 201]. For the case where the recovery paths are computed in real time after a failure is detected and localized, the reader is referred to [64, 138]. A brief discussion on restoration techniques was also included in Chapter 2.

More specifically, this book examines in detail the case of DBPP and SBPP in networks with shared risk groups (SRGs). Both concepts were briefly discussed in Chapter 2. In the case of DBPP, the working and backup paths are SRG-disjoint and no sharing among the redundant capacity is possible. The provisioning algorithm for this architecture computes and establishes simultaneously the primaries and their SRG-disjoint secondary paths. During normal operation mode, both paths carry the optical signal and the egress node selects the best copy of the two. This is the fastest protection scheme, since for every lightpath one device is responsible for all the necessary failure detection and recovery functions. But it is also the most exigent in terms of resource consumption.

In the case of SBPP, two working paths can now share capacity on a backup path if they are SRG-disjoint. In the sections that follow, the reader will go through a description of SRGs, the different approaches used for node vs SRG failures, as well as several performance results for a host of different network architectures.

3.3.1 Shared Link Risk Groups

Failures of multiple optical channels are usually due to fiber or cable cuts. Consider the six-node optical network of Figure 3.3 (see below). Each cylinder in the figure represents a conduit. Optical channels across the two links connecting two distinct pairs of nodes traverse the same conduit. If the conduit, the cables, and the fibers it contains are accidentally severed, all the optical channels inside the conduit fail (see Figure 3.2). Furthermore, one fiber may traverse one or more conduits, and cables in one conduit may have varied origins/destinations.

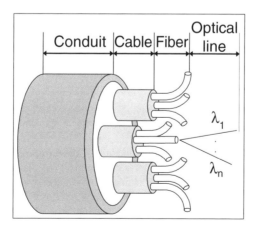

Figure 3.2: Representation of a conduit.

The concept of SRG expresses the risk relationship that associates all the optical channels with a single failure [250, 290]. An SRG may consist of all the optical channels in a single fiber, of the optical channels through all the fibers wrapped in the same cable, or of all the optical channels traversing the same conduit, or the same equipment.

There seems to be a disconnect in the industry from a group of engineers that assumes a very large number of SRGs in the network by considering every fiber as an SRG for all the lightpaths that traverse it, and a group of engineers that assumes a small number of SRGs by considering entire cables (or even conduits) as the SRGs. Clearly, a network designer or a network operator can make any one of these assumptions depending on the risk of failures exhibited by the network components. Both of these approaches have their advantages and disadvantages. Note that the latter case encompasses the former as well, since if a cable is an SRG for the lightpaths that traverse it then the fibers that are included in this cable are also SRGs for these lightpaths. The latter will provide for enhanced survivability, as all lightpaths will be routed via diverse cables, thus ensuring that any cable failure will not result in the loss of a connection. In addition, this approach assumes a small number of connections and it makes the problem of finding and managing the SRGs relatively simple for the network designer or the network operator. The drawback of this approach is that it may sometimes be difficult to find diverse routes for two connections that use different fibers but traverse the same cables. On the other hand, the former approach can find diverse routes rather easily, as it can use two different fibers that belong to the same cable. However, the former approach implies that the network operator will have to manage a very large number of SRGs and this technique may lead to lost connections if a cable failure occurs and the primary and secondary paths of a connection use two different fibers in the same cable for diverse routing. The same reasoning applies if the fibers in the arguments above are replaced by cables and the cables are replaced by conduits and so on.

For obvious reasons, a network topological view alone does not encompass the notion of SRGs. With the exception of the default case, there exists no simple way to automatically generate this information. The network operator must provide it. There were some attempts made for SRG autodiscovery such as the location-based technique introduced in [273, 274]. In that case, the authors proposed that all components at risk (active and passive) should be assigned unique network identifiers, certain components should be equipped with location-finding devices (such as Global Positioning Systems (GPS)), and for other components, the location information should be manually entered into an associated database. This way, the (possibly) flawed, manually maintained databases can be substituted by a more automated and reliable system used to support SRG management.

Since a fiber may run through several conduits, an optical channel may belong to several SRGs. The default case is clearly the one where all the optical channels between one node pair belong exclusively to one SRG. The routing algorithms used to calculate backup paths have to exploit the SRG maps to discover SRG-diverse routes so that after any conduit is cut, there is always at least one viable route remaining for recovering the failure [145, 205, 228, 234, 290, 316, 332]. For instance, in Figure 3.3, the subtraction of any SRG and the optical channels that traverse it, affects at most one of the two shown routes from A to B.

The term SRG is used in this book in the generic sense to signify the risk interdependencies. The terms SRLG (shared risk link group), SRNG (shared risk node group) and SREG (shared risk equipment group) are more specific, signifying the shared network components in these interdependencies (i.e., links, nodes or equipment).

All SRGs can be expressed as one or a combination of three possible primary types. These are described in Figure 3.4 [107]. The default and most conventional type, is type (a) in the figure, which associates an optical channel risk failure with a fiber cut. Another type of SRG very likely to be encountered is type (b) (also called *fork* SRG). This type is typical of fibers terminating at a switch and sharing the same conduit into the office; that is, a conduit cut would affect all the optical channels terminating at the switch. Types (a) and (b) can be characterized in a graph representation as pictured by graphs (a′) and (b′) in Figure 3.4. For instance, the removal of the edge in (a′), or the middle

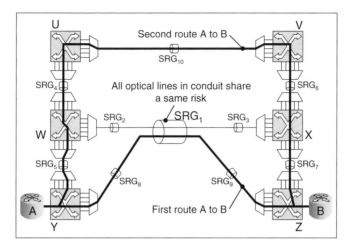

Figure 3.3: Shared risk groups. (From [107], Figure 3a. Reproduced by permission of © 2003 The International Society for Optical Engineering.)

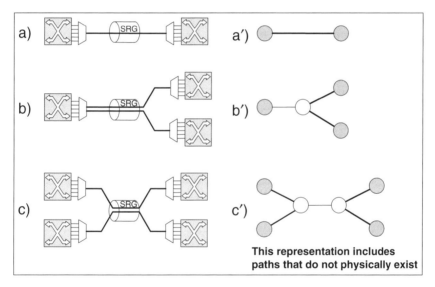

Figure 3.4: SRG classification. (From [107], Figure 3b. Reproduced by permission of © 2003 The International Society for Optical Engineering.)

node in (b′), disconnects all the nodes which is tantamount to an SRG failure in each case. Using these elementary transformations it is possible to model the network as a graph onto which established shortest-path operations can be applied. Type (c) SRG is the most difficult kind to model and provide diverse routing for. It occurs in a few instances, such as for example fibers from many origins and destinations routed into a single submarine conduit, or dense metropolitan areas. Contrary to types (a) and (b), there is no convenient way to graphically represent type (c) SRGs and their presence can increase dramatically the complexity of the SRG-diverse routing problem. A naive representation of the type (c) SRG would be to present it as graph (c′) in Figure 3.4. Such a representation, however, introduces additional paths not present in the original network topology, which could lead us to routing computations that are not physically feasible. Thus, if such a representation is made it has to be treated appropriately. This means that the network architect has to recognize that there are some additional paths that do not physically exist and add this constraint in the corresponding routing algorithms.

3.3.2 Shared Node Risk Groups

Two types of protection are considered: (1) SRG failure resilient and (2) node failure resilient networks. Resilience against SRG failures is achieved by way of path diversity, as illustrated in Figure 3.5(a). Some level of node failure protection can be realized by way of a redundant switch fabric. Note that this architecture does not recover from severe events, such as electrical fires or floods, that affect both switching fabrics. If recovery against this type of failure is also desired, it is necessary to provision routes that are SRG- and node-disjoint, as shown in the example of Figure 3.5(b). However, node-diverse paths consume more resources than the less conservative SRG-diverse scheme pictured in Figure 3.5(a).

3.3.3 Shared Equipment Risk Groups

It is a well-known fact that the demand for bandwidth has been growing at a steady pace, requiring denser optical switches capable of packing more bandwidth into congested central office spaces. Higher density is in part achieved with the use of circuit packs,[5] with multiple pluggable modules, which minimize hardware redundancy, footprint and power consumption by sharing functionalities among multiple optical interfaces. The optimum number of pluggable optical interfaces sharing a circuit pack depends on two competing factors. On one hand the port density of the optical switch is proportional to the density of the circuit pack itself, and it is thus desirable to increase the number of interfaces per circuit pack in order to maximize the port density of the switch. On the other hand, as functionality migrates from the interfaces onto the circuit packs, the interfaces sharing the circuit pack must have some level of commonality, such as bit rates, or signal format (e.g., TDM, SONET, Ethernet). As a consequence, it is often necessary to equip a switch with multiple types of circuit packs to manage the large variety of interfaces. This is particularly true on the access side of the network, where the traffic emerges in various forms from the customer premises and is aggregated into a format suitable for transport through the core network. Under these circumstances, the use of circuit packs leads to capacity fragmentation, known as derating, and a less efficient utilization of the switch capacity if all the ports of the modules are not occupied. This is further exacerbated by the fact

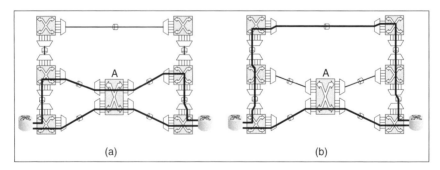

(a)   (b)

Figure 3.5: (a) SRG-diverse paths (b) Node-diverse paths. (From [107], Figure 4. Reproduced by permission of © 2003 The International Society for Optical Engineering.)

---

[5]OEO switches are assumed in this case, as described in Chapter 1.

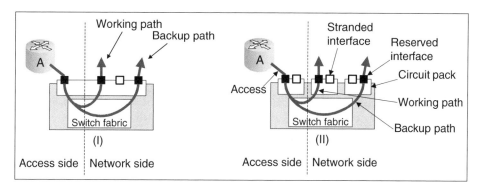

Figure 3.6: Ingress of an end-to-end protected circuit using various circuit pack configurations with increasing level of protection. (From [53], Figure 3. Reproduced by permission of © 2006 The Optical Society of America.)

that each circuit pack constitutes a single point of failure, whereby all the interfaces it contains could be simultaneously affected if the module happens to fail, or is decommissioned for maintenance.

Clearly, diversity requirements should apply to circuit packs as well. As illustrated in Figure 3.6, a protected lightpath (circuit) can have all its network ports plugged on the same circuit pack, or use a separate module for each of its access and network ports. Although this is not shown in the figure, the network operator can further have the choice to protect the access link using $(1 + 1)$ or $(1 : N)$ protection on the same module or on redundant modules. The first configuration ignores the circuit pack diversity constraint, and the circuit is susceptible to the failure of a single module. The second configuration is more robust, since the circuit is recovered upon the failure of a single circuit pack. However, this configuration also requires the largest number of circuit packs, and is hence likely to be penalized by a lower overall switch utilization than other configurations. The impact of the diversity and sharing constraints is further illustrated in Figure 3.7. The figure depicts two SBPP demands routed on an optical network between a pair of nodes A and Z. Circles are used in the figure to indicate groups of ports that can be collocated on a same circuit pack without violating the diversity and sharing constraints on protection against circuit pack failures.

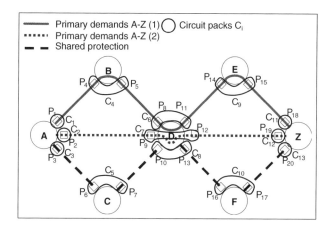

Figure 3.7: Primary and backup paths of an SBPP demand, and circuit pack configuration. (From [53], Figure 4. Reproduced by permission of © 2006 The Optical Society of America.)

Circuit pack diversity routing algorithms as well as experiments with different sizes of circuit packs assuming an SBPP architecture are analyzed in detail in Chapter 4.

3.4 Centralized vs Distributed Routing

Calculation of the protection or restoration routes (route and channel selection for each) can take place utilizing either one controller with complete knowledge of the network parameters (centralized approach) or utilizing the controllers at every node in the network (distributed approach). Figure 3.8 shows an example of the centralized vs distributed routing process. The sections that follow discuss both the centralized and distributed routing approaches [24] for protection/restoration and try to identify (qualitatively and via simulation) the strengths and weaknesses of each technique.

3.4.1 Centralized Routing

In the centralized approach, the joint problem of route computation and channel assignment for a light-path is solved by a Centralized Management System (CMS) that has access to the complete network state of the optical cross-connects, including the topology and lightpath databases. In centralized protection or restoration schemes, the central controller calculates the protection/restoration routes prior to or after the failure event respectively. In order for the central controller to be able to identify the best possible route for failure recovery, it should possess all the relevant information associated with the network, and with each network link and network node. This information may include the network topology (including the distances between network nodes), the working and redundant capacity available on each link (real-time updates), the traffic currently on the network (connectivity maps), future traffic demands (if known), possible node/link avoidance or node/link inclusion, etc.

If a restoration technique is implemented, after the failure event, the central controller is notified of the failure, and based on the state of the network at the time of the failure, it calculates the most efficient restoration path. On the other hand, if a precomputed protection technique is implemented, after the failure event, the central controller goes through a list of predefined backup paths for every possible failure scenario and determines the appropriate backup path for the specific failure event. For both cases, after the central controller has obtained information on the recovery path, it then notifies all nodes on that path to execute the necessary cross-connects. In centralized protection/restoration

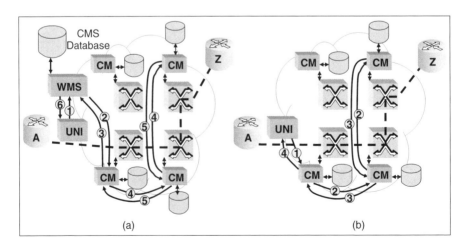

Figure 3.8: (a) Centralized routing, (b) Distributed routing.

techniques, the time it takes to recover from the failure is significant, as the complete network information is being used to calculate the best possible recovery path. This time depends mainly on the network size and the control architecture used in the network. However, exactly because the complete network information, based on the current state of the network, is used to calculate the recovery path, centralized approaches result in much more efficient capacity utilization, compared to their distributed counterparts [162, 251].

An example of centralized provisioning is shown in Figure 3.8(a), where a request for a lightpath is established through the UNI or the CMS. The CMS computes the route, assigns the optical channels along the route, and sends a request to establish the lightpath to the ingress Control Module (CM) (controller associated with each node) of the route. The message sent by the CMS to the CM contains the description of the route and the port numbers of each optical channel. The ingress CM configures the associated OXC to create a new connection for the lightpath in accordance with the information provided by the CMS, or returns an error message with the updated state of the switch to the CMS if it cannot create the connection. If the connection can be created, the CM forwards the request to the next CM on the route and waits for lightpath set-up confirmation from this CM. The lightpath is established and the CMS returns *success* to the UNI if all CMs along the route successfully complete the cross-connection.

Approaches to solve the centralized routing problem when SRGs can assume general configuration include (a) enumerative approaches based on enumerating the k-shortest paths for primary and backup paths and selecting the primary and backup paths that meet the delay constraints, and (b) search heuristics that search for the feasible primary and backup paths that meet the delay constraints. It was shown that for network topologies of approximately 50 nodes, enumerative approaches work well, and yield optimal solutions for small values of k ($k = 10$). Running time for enumerative approaches is of the order $O(kN^2)$. For example, for a 50-node network with average degree 3, these approaches run in O(seconds) [251].

An example of a centralized online route-provisioning algorithm that protects against both single node and SRG failures is outlined in Table 3.2. What is generally given in these cases is a mesh optical network topology with programmable OXCs (nodes) interconnected by bidirectional fiber pairs (links). This topology is modeled as a directed graph with edges representing the network links and vertices representing the network nodes. The demand arrival is presented by $K = \{(u_1, v_1, k_1, g_1), \ldots, (u_m, v_m, k_m, g_m)\}$ ordered according to time of occurrence. A quadruplet in K indicates the node pair (u_i, v_i), the requested traffic amount k_i, and a protection descriptor g_i (more about this later in this section). For the sake of simplicity, and without loss of generality, assume in the remainder of this discussion that $k_i = 1$ unit of the rate supported by the optical switch (e.g., one unit of OC-48). In addition, OXCs are assumed to have unlimited number of ports and are nonblocking; and each link e_i supports up to a predetermined number c_i of channels of some prescribed rate (link capacity). Also, given is an SRG map $S = \{s_1, \ldots, s_m\}$ that delimits subsets of optical lines sharing same risk characteristics.

The objective of the routing algorithm would be to route the demands online subject to:

- Capacity constraints, and

- Provision for protection: demands by default are guaranteed to survive any single SRG failure. In addition, the protection requirement of each demand may include recovery from an arbitrary subset of node failure scenarios. The protection descriptor g for a demand provides the necessary information on the type of protection that is desired and the list of critical nodes whose failures must not disrupt the demand. The motivation for permitting individual lists of critical nodes results from the fact that some demands should exist only if specific sets of nodes are active and working. It is precisely this group of nodes that g_i refers to.

In the description of the algorithm, an exhaustive knowledge of the state of the network is assumed. Rules of sharing formulated in Section 3.3.1 are used. The algorithm is executed whenever a request

Table 3.2: Example of a centralized online route-provisioning algorithm that protects against both single-node and SRG failures

---

Definitions:

- Given the topology, represent every optical switch as a vertex, and every bidirectional fiber-link as an edge. Let $G(V,E)$ be the resulting graph with vertex set V, and edge set E. In finer details, an edge actually consists of two parallel arcs oriented to opposite directions between the same pair of vertices.

Algorithm:

1. Compute K-shortest paths. Label the paths from p_1 to p_k.

2. $S \leftarrow \{\}$.

3. For each p_i ($1 \leq i \leq k$), compute a backup path as follows:

   - Temporarily set weight on edges:
     * For each edge that belongs to an SRG on p_i, set weight to infinity.
     * Remove nodes on the critical set g.
     * For each edge that has channels shareable with p_i, as determined in Section 4.3.2.3, set weight to 0.
     * For each edge that is not shareable with p_i, but has idle channels, set weight to link cost.
     * Otherwise set weight to infinity.
   - Compute a shortest path, label the path q_i.
   - $S \leftarrow S \bigcup (p_i, q_i)$
   - Reset weights to their previous values.

4. In S, select the pair (p_i, q_i) that induces the minimum cumulative weight. Break ties with the pair that has the shortest primary.

5. Determine channels on p_i and q_i (in both directions).

6. Update the Channel Sharing Database: newly assigned channels on q_i are marked as nonshareable with any SRG or critical vertex of p_i.

7. Return (p_i, q_i).

---

to accommodate a demand (u, v, k, g) is received. In the description of the algorithm in Table 3.2, K-shortest paths are calculated. The reader is referred to Chapter 6 for more details on algorithms on how to accomplish this.

3.4.2 Distributed Routing

Distributed protection or restoration techniques do not require a central controller with a centralized view of the entire network, with up-to-date information on the state of the network when a failure occurs, and no network node has a global network description. Rather, the local controllers that are present at every network node are the ones that facilitate the recovery route computation and the implementation of the recovery process. The reader is referred to Chapter 9 for a complete description of path routing with partial information.

In the distributed routing case, the local controllers at every network node store information mostly about their own links (e.g., working and redundant capacity on these links) and about their own and neighboring nodes (e.g., the nodes that they are connected to, the connections through their node) [171]. If a precomputed approach is utilized, the local controllers also store some information on possible backup path routes that will be used after a failure event. When a failure occurs, these nodes can then initiate the recovery process and execute preset algorithms to establish signaling communication among network nodes as well as execute the necessary cross-connects to effect the rerouting of the affected signal path. If a real-time restoration approach is utilized, after the failure event, these local controllers are used to initiate restoration signaling between the nodes so as to update the nodes with relevant topology and available capacities, and to compute and establish the new routes.

For networks of large size and with a large number of active connections, the distributed techniques will perform better than their centralized counterparts in terms of recovery time (scalability is an issue for any centralized technique) [251]. This advantage will be less evident however for relatively small networks. The reader should also note that in some cases, the recovery technique utilized could be a combination of a centralized and a distributed approach. For example, the SBPP technique as described throughout the book follows precisely that approach: the route computation is done in a centralized manner, while the signaling and execution of cross-connects are done in a distributed manner.

In distributed provisioning, the UNI request is sent directly to the ingress CM, which acts as a substitute for the CMS. In the distributed approach, lightpaths are routed by the CM of the ingress OXC using its local databases. The local database at the switch contains the summarized topology disseminated by a link-state protocol such as OSPF, as well as those provisioned demands whose primary or backup paths traverse that switch. Shareability of optical channels for SBPP connections is not known at the time of route computation. Routing is therefore performed first and is in the most part oblivious to the network-wide configuration of the lightpaths; channels are assigned next. The CM then executes the cross-connection, submits the request to set up a lightpath to the next CM along the route and waits for lightpath set-up confirmation from this CM. Figures 3.8(b) and 3.9 show the successive steps for a distributed provisioning approach.

In this distributed approach the ingress OXC computes the primary and backup paths as follows: it first enumerates the k-shortest paths for primary paths that satisfy the delay constraint and then selects the *best-effort* shared backup paths that meet the delay and recovery time constraints. If the

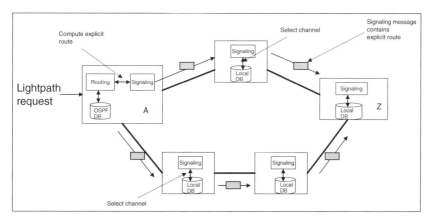

Figure 3.9: Distributed route computation and provisioning.

backup path signaling fails to find available channels, then the signaling cranks back to the ingress OXC with information about the optical line on which a channel could not be found. The ingress OXC then retries by computing a different backup path by excluding the optical line which previously blocked the backup path.

Delegating routing operations to the CMs (distributed routing) clearly enhances network scalability. The downside of distributed routing is that it is possible that the route is suboptimal or even cannot satisfy all the constraints (although some techniques can be employed to estimate the state of the network using appropriate partial information [51]). Such a solution will lead to increased network capacity requirements and higher network cost [250]. Multiple routing attempts may also be necessary before a feasible path is found.

The reader should note that, as mentioned before, the strategy employed in routing and that used in the recovery mechanism can be independent. It is possible, for instance, to provision a lightpath in a centralized manner, while relying on a distributed mechanism to protect it.

An example of a distributed online route-provisioning algorithm that protects against both single node and SRG failures is outlined in Table 3.3. The same assumptions as in the previous section, which described a centralized online route-provisioning algorithm, hold, except that each node now maintains only a local Channel Sharing Database, in addition to the OSPF topology database. The route computation is performed at the ingress switch. The algorithm is again executed whenever a request to accommodate a demand (u, v, k, g) is received. Again, given the topology, every optical switch is represented as a vertex, and every bidirectional fiber-link as an edge. $G(V, E)$ is the resulting graph with vertex set V, and edge set E. In finer details, an edge actually consists of two parallel arcs oriented to opposite directions between the same pair of vertices.

Both the centralized and distributed algorithms outlined can easily allow for incorporating constraints for the demand such as propagation delay on primary/backup lightpath, or recovery time for the lightpath. The reader is referred to Chapter 6 for a complete analysis of routing heuristics.

3.4.3   Centralized vs Distributed Routing Performance Results

The first set of experiments simulates a backbone network with 30 nodes, average node degree 3.1 and a full-mesh demand of 30×29 lightpaths for both centralized and distributed route computation. For both cases SBPP is implemented. The results shown in Figure 3.10 demonstrate that distributed route computation in this test case requires approximately 12% more total capacity compared to centralized route computation. It is important to note that it is not so much the distributed approach as the lack of complete information that causes the difference here. This is the case as the distributed approach implies limited information at the place the routing computation is carried out. Offline reoptimization of lightpath backup routes, which is discussed in detail in Chapter 10, is one way to recover the capacity penalty in the case of distributed route computation. Additional results in Figure 3.11 show that, on the average, backup paths are three hops longer for centralized route computation over distributed route computation.

As mentioned above, crankbacks are also utilized if the backup path signaling fails to find an available channel. Figure 3.12 shows the impact of crankbacks for distributed route computation. Crankbacks reduce time to first blocking and can reduce total blocking in the network. Furthermore, two to three crankbacks are sufficient in the network to find an appropriate backup path.

The next set of experiments compares centralized and distributed routing. The N_{17} and N_{100} networks, with 17 and 100 nodes respectively, are used as in the case of the comparison between DBPP and SBPP (Section 4.4), but with different order of demands. The assumption here is that distributed routing has access to topological information and link utilization, but does not have information on existing lightpaths, and thus cannot derive lightpath compatibility information. The reserved channels are thus assigned after the routes are computed – as opposed to centralized routing, where routes can be computed to maximize the sharing and minimize the allocation of new optical channels on primary and backup paths.

Table 3.3: Example of a distributed online route-provisioning algorithm that protects against both single-node and SRG failures

---

Algorithm:

1. Compute K-shortest paths. Label the paths from p_1 to p_k.

2. $S \leftarrow \{\}$.

3. For each p_i $(1 \leq i \leq k)$, compute a backup path as follows:

   - Temporarily set weight on edges:

     * For each edge that belongs to an SRG on p_i, set weight to infinity.
     * Remove nodes on the critical set g.
     * For each edge that has channels shareable with p_i, as determined in Section 4.3.2.3, set weight to 0.
     * For each edge that is not shareable with p_i, but has idle channels, set weight to link cost.
     * Otherwise set weight to infinity.

   - Compute a shortest path, label the path q_i.

   - $S \leftarrow S \bigcup (p_i, q_i)$

   - Reset weights to their previous values.

4. In S, select pair (p_i, q_i) with the minimum cumulative weight. Break ties with the pair that has the shortest primary.

5. Use a signaling protocol to reserve channels on p_i.

   - If the channel assignment succeeds, let $p = p_i$ and $q = q_i$.
   - Otherwise, prune the first unsuccessful edge on p_i, and go back to Step 1.

6. Repeat a limited number of times (number of attempts) for setting up the backup route.

   - Use a signaling protocol to reserve channels on q. Sharing is applied whenever possible in accordance to the rules of sharing stipulated in Section 3.3.1.
   - If the channel assignment succeeds, return SUCCESS.
   - Otherwise, set weight of the first unsuccessful edge on q to infinity, and recompute the backup path q, that is diverse from the already established primary path p.

     * For each edge that belongs to an SRG on p, set weight to infinity.
     * Remove nodes on the critical set g.
     * For each edge that has available channels shareable with p_i, set weight to 0.
     * For each edge that is not shareable with p_i, but has idle channels, set weight to link cost.
     * Otherwise set weight to infinity.
     * Compute the shortest path.

7. If number of attempts exceeds the prefixed limit, hang up p (free reserved channels) and block the lightpath.

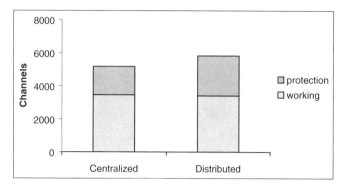

Figure 3.10: Comparison of centralized vs distributed routing in terms of capacity requirements for a 30-node network with average node degree 3.1 and a full-mesh demand.

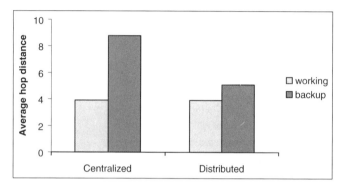

Figure 3.11: Comparison of centralized vs distributed routing in terms of the average number of hops for the backup paths for a 30-node network with average node degree 3.1 and a full-mesh demand.

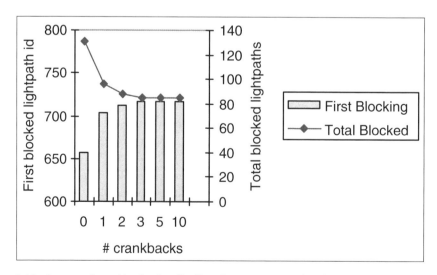

Figure 3.12: Impact of crankbacks for distributed route computation for a 30-node network with average node degree 3.1 and a full-mesh demand.

Results for the N_{17} network are shown in Figure 3.13. Experiments indicate that distributed routing of SBPP connections incurs a capacity penalty of 12 to 17% over centralized routing. Similar results for the N_{100} network can be observed in Figure 3.14 [107].

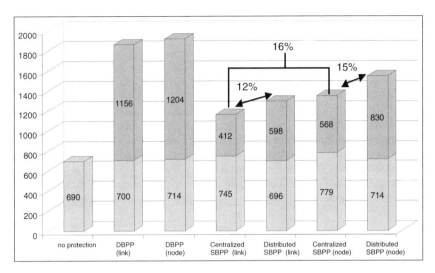

Figure 3.13: Comparison of centralized vs distributed routing for different protection architectures (17-node network). (From [107], Figure 13. Reproduced by permission of © 2003 The International Society for Optical Engineering.)

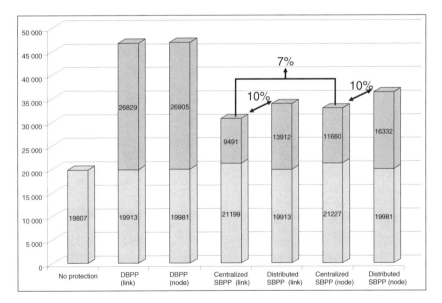

Figure 3.14: Comparison of centralized vs distributed routing for different protection architectures (100-node network). (From [107], Figure 14. Reproduced by permission of © 2003 The International Society for Optical Engineering.)

3.5 Conclusion

In this chapter we have further classified the protection and restoration architectures in terms of link vs path, precomputed recovery route computation vs computation of the recovery route in real time, and centralized vs distributed route computation. This chapter introduced the concept of shared risk groups (SRGs), demonstrated three different types of SRGs and showed how these SRG types can be modeled in the network.

We demonstrated that there are different diversity requirements for different types of failures (node vs SRG failure protection) and also noted that diversity requirements apply to circuit packs as well. We also discussed shared segment protection, an approach that falls between link and path-based protection, offers comparable performance in terms of capacity requirements to the path-based technique, and is flexible enough to avoid problems of finding SRG-disjoint paths in trap topologies.

Finally, we focused on the centralized vs distributed routing problem and showed that the centralized routing approach typically requires less capacity than the distributed approach. However, centralized routing requires a higher (on the average) number of hops for its backup paths, as it is trying to maximize sharing as much as possible while having the complete topology information. If crankbacks are used when the backup path signaling fails to find an available channel for distributed route computation, we further showed that two to three crankbacks are sufficient to find an appropriate backup path.

Chapter 4

Path Routing and Protection

4.1   Introduction

In this chapter, and in the rest of the book, we mostly concern ourselves with routing and path-based protection of connections under the preplanned failure-independent model captured in Table 3.1 as categories 1 and 2 [93, 99, 107, 179, 201]. We will first present a framework for routing-path protected connections in a mesh network, describing routing attributes of importance. We then present a framework for describing several path protection schemes, in particular Dedicated Backup Path Protection (DBPP) and Shared Backup Path protection (SBPP). In addition, we will introduce and discuss additional types of connections, such as unprotected connections, preemptible connections, connections with multiple protection paths or with relaxed protection guarantees. We will also cover the case of connections that leverage the concept of dual-homing to protect against failures in the local loop, or access to the core optical mesh network. In subsequent chapters, we will analyze the complexity of such routing problems (Chapter 5) and propose a number of routing algorithms, exact or heuristic, to solve these routing problems (Chapter 6).

The chapter is structured as follows. Section 4.2 deals with the routing of path-protected connections. Section 4.3 presents the two primary types of protected connections, namely DBPP and SBPP connections, as well as a number of additional types of protections. Sections 4.4 and 4.5 present a number of performance results in terms of overall capacity utilization in the networks, as well as recovery time. In Section 4.6, we discuss some of the trade-offs that influence both the capacity utilization and the recovery time. Finally, we draw conclusions in Section 4.7.

4.2   Routing in Path-Protected Mesh Networks

Routing in a multi-layered architecture is preferably done separately in the logical and in the optical layers [106]. It is the latter that is addressed here, in the context of an opaque network built with OEO switches [108].

The procedure to route a lightpath consists of two tasks: (1) route selection, and (2) channel selection. Route selection involves computation of the primary and backup paths from the ingress port to the egress port across the mesh optical network. Channel selection deals with selecting individual optical channels along the primary and backup routes. Channel selection is obviously done at provisioning time for the primary and backup paths of a DBPP lightpath. In the case of the backup

Path Routing in Mesh Optical Networks   Eric Bouillet, Georgios Ellinas,
Jean-François Labourdette, Ramu Ramamurthy   © 2007 John Wiley & Sons, Ltd

path of a SBPP lightpath, two modes are possible, pooled backup channels and preassigned backup channels as described in Section 4.3.2.1.

The problems of selecting a route together with selecting channels on the route are closely coupled and if an optimal solution is sought both problems should be solved simultaneously. When solving the route computation problem, several attributes and metrics need to be considered. Depending on the allotted budget, the desired protection QoS (dedicated vs shared backup path, link only vs link and node failure protection), and the desired performance QoS (traffic latency, recovery time), each attribute and metric either enters as a parameter in the algorithm's objective function to be minimized, or is used as a constraint to eliminate solutions that do not meet practical limits. Some of these metrics are:

- **Cost:** The use of optical channels on both the primary and backup paths (for protected connections) entails a cost.[1] It is henceforth important to ensure that the cumulative cost does not exceed the client's budget. The cost associated with optical channels will differ based on their current use, so that for example, the cost of using a channel on a shared protected backup path already part of the pool of shared backup channels should be less than the cost of using a brand new channel as a shared backup channel. Channel cost models and their impact on path routing are investigated in Chapter 7.

- **Bit-rate:** Each optical channel is set to a predetermined bit-rate (e.g., OC-3, OC-48, etc.) The bit-rate of all the selected optical channels along the route must meet the lightpath's bandwidth prescribed by the client layer through the Network Management System (NMS) or through UNI interaction with a client node.

- **Resiliency/protection type:** The quality of service resilience is judged based on the level of protection offered with the service against a range of failures (e.g., dedicated vs shared backup path protection; link vs node and link failure protection, single failure vs multiple failures). The client expects certain guarantees on the robustness of the connection. However, the optical carrier can reduce but not eliminate every risk, such as fiber cuts. It is therefore important to provision backup capacity on alternate routes where services can be recovered if failures occur on the primary lightpath. Based on the type of protection requested, additional attributes may be necessary. These were addressed in general in Chapter 3 and will be discussed later in this chapter in more detail.

- **Service availability:** Service availability, typically measured in downtime minutes per year, could be used as a performance metric, which could determine the type of protection (dedicated vs backup path protection, link vs link-and-node protection) required to achieve the required availability performance. Service availability is covered in Chapter 12, and will not be considered explicitly in the discussion on routing algorithms as it is extremely difficult to model in a way that could be accurately incorporated in the design of routing algorithms.

- **Primary path latency:** The latency on the primary path is the latency experienced by the traffic on the connection in normal (nonfailure) mode, and is composed of propagation delay as well as processing delay at the nodes traversed by the primary path. Latency is a cumulative metric that requires a link length or link delay attribute.

- **Backup path latency:** The latency on the backup path is the latency experienced by the traffic on the connection in failure mode after the traffic has been switched to the backup path, and is composed of propagation delay as well as processing delay at the nodes traversed by the backup path.

---

[1]In addition, there is another overall hidden cost of capacity used, or conversely a premium for efficiency in use of capacity. Efficient use of capacity allows the operator to better handle unexpected demands or traffic patterns, and to postpone the need for reoptimization or capacity expansion.

- **Recovery time:** The recovery time is the time experienced by a connection to be reestablished on the backup path following a failure that affected the primary path. DBPP connections typically would have lower recovery time than SBPP connections, but at a higher cost since the protection path is dedicated to a connection. Among SBPP connections, the recovery time experienced during a failure event can be modeled and analyzed (see Sections 4.5 and 11.8) and the resulting information can be used to influence the routing of such connections. The recovery time depends on the recovery architecture and implementation and on the characteristics of the path selected in the routing algorithm such as mile-length of the path, and average load in the nodes along the path. These are determining factors in the route computation algorithm.

The objective of the routing algorithm is either to preserve spare capacity by using a minimum number of optical channels, or to find the solution that incurs the minimum cost. Therefore, among the link attributes that are necessary to achieve routing operations, are (1) the cost per optical channel, (2) their bit-rates, (3) the protection type to achieve a certain level of resiliency. Path latency and recovery times are more difficult to directly incorporate in the routing objectives and are usually handled as constraints in the routing framework.

Lastly, available information about the state of the network has a direct impact on the design of routing algorithms. The view of the network varies depending on where the routes are computed. It can be global with the maximum knowledge about network state and link attributes if the computation is centralized, or local with very sparse knowledge if it is distributed across the network. In the latter case it may be necessary to produce an educated guess of the route with whatever information is available and defer feasibility and optimality questions – whether all requirements are met – to the moment the lightpath is effectively established through the optical switches. In this chapter, we assume that routing computation is done with access to the complete network information. Route computation when only partial information is available is discussed in detail in Chapter 9 as well as in [51, 200].

It is also important to remember that the routing problem addressed in this book is online routing in an opaque network architecture that utilizes OEO switches [108], where wavelength conversion is a byproduct of the switch architecture. Such architecture was described in Chapter 1. We do not address routing in transparent networks where optical impairments such as loss and dispersion accumulate in complicated and nonlinear ways to impact on the routing of lightpaths. We refer our readers to [290] for a discussion on that subject.

4.3   Protection in Path-Protected Mesh Networks

In path or end-to-end protection, the ingress and egress nodes of the failed optical connection attempt to recover the signal on a predefined backup path, which is SRG-disjoint, or diverse,[2] from the primary path [107, 187]. Path diversity or disjointness guarantees that primary and backup lightpaths will not simultaneously succumb to a single failure. The book by Bhandari [40] is devoted to and provides a comprehensive treatment of diverse routing problems. In the recovery architecture we consider, backup paths are provisioned along with their corresponding primary paths, and thus the recovery or protection switching from primary to backup path does not involve further real-time path computations.

Typically, the detection of signal failure or signal degradation is done at the port/switch adjacent to the failure (e.g., fiber cut), and the failure condition is communicated or signaled to the endpoints of the connection. As described in Chapter 2, signaling can be carried out in multiple ways, through a separate signaling architecture, or in-band using overhead bytes in the SONET/SDH structure of the connection. See also [38] for additional information on the signaling architectures for optical mesh networks.

---

[2]*Diversity* is a common industry term that in this context actually means disjoint routing, not simply distinct or diverse.

Another advantage of path protection is that the processing is distributed among ingress and egress nodes of all the lightpaths involved in the failure, compared to local span protection where a comparable amount of processing is executed by a smaller set of nodes adjacent to the failure. In the following we will only consider cases where the protection path is failure-independent and is thus the same for all types of failures. By way of this restriction, the protection paths may be computed and assigned before failure occurrence. As introduced in Chapters 2 and 3, there are two main types of path protection: (1) Dedicated Backup Path Protection (DBPP), and (2) Shared Backup Path Protection (SBPP). Variations on these two main types are also considered in the subsequent sections.

4.3.1   Dedicated Backup Path-Protected Lightpaths

4.3.1.1   Protection against Single Link-SRG Failures

Dedicated backup path protection is illustrated in Figure 4.1. The network consists of four logical nodes (A to D) and two demands (AB and CD) accommodated across an eight-node optical network (S to Z). The provisioning algorithm for this architecture computes and establishes simultaneously the primaries and their link-SRG-disjoint protection paths. During normal operation mode, both paths carry the optical signal and the egress node selects the best copy of the two. In the example of Figure 4.1, all the optical channels on primary and secondary paths are active. In particular, the configuration reserves two optical channels between nodes S and T for protection. This is the fastest protection scheme, since for every lightpath one device is responsible for all the necessary failure detection and protection switching functions. But it is also the most expensive in terms of resource consumption.

The problem of finding SRG-diverse routes for dedicated path-based protection, in order to accommodate single SRG failures, is trivial if SRGs are of type (a) or (b) (see Figure 3.4), or a combination

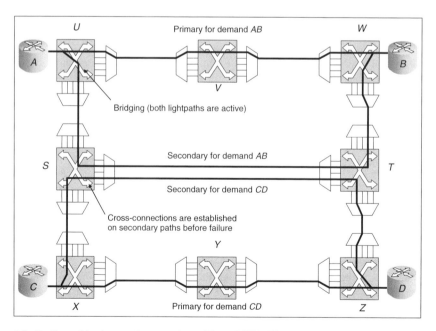

Figure 4.1: Dedicated backup path protection. (From [107], Figure 5. Reproduced by permission of © 2003 The International Society for Optical Engineering.)

of both, since they can be easily represented in a graph model as shown in Figure 3.4. The goal in such an approach would be to minimize the total cost[3] for the optical lines used in the primary and backup paths, subject to bit propagation delay constraints on primary/backup paths. To solve such a problem, the network is modeled as a directed graph, with each directed edge of the graph corresponding to an optical line in the optical network, and those lines have a weight and a delay metric. When the SRGs assume default configurations, and there are no delay constraints, there exist optimal algorithms (in terms of minimizing the capacity required), such as Suurballe's algorithm [293] explained in Chapter 6, to solve this problem by finding diverse routes. When SRGs can assume general configuration (and even without any delay constraints), including the case of type (c) SRGs (see Figure 3.4), finding SRG-disjoint routes is a provably NP-complete problem, and pseudo-optimal solutions must be obtained using enumerative approaches. The proof of the NP-completeness of this problem is given in the appendix of Chapter 5. Essentially, the difficulty of SRG-diverse routing arises because the architecture allows SRGs to be defined in arbitrary and impractical ways which forces an algorithm to enumerate (a potentially exponential number of) paths in the worst case (unless P = NP). The problem is also NP-complete if the selected routes must respect a set of independent constraints, such as maximum round trip delay (or alternatively maximum path length expressed in geographical distance units).

4.3.1.2 Protection against Single Node and Link-SRG Failures

If protection against node failure is also desired, then primary and backup paths must be node-disjoint in addition to link-SRG-disjoint. As explained earlier, node protection may require more protection capacity. However, experiments indicate that the two types of protection use a comparable amount of capacity. The problem of finding node-diverse routes is equivalent to the problem of link-diverse path routing. The same algorithm is capable of solving either problem, using the graph transformation pictured in Figure 4.2 to represent nodes as directed links, by introducing an in-node (shown as ⊖) and an out-node (shown as ⊕) [107]. Clearly, if a pair of paths is edge-disjoint in Figure 4.2(b), then it is also node-disjoint in Figure 4.2(a). In this example there is no edge-disjoint path from *A* to *Z* in the graph representation, and thus no node-disjoint pair of paths in the network.

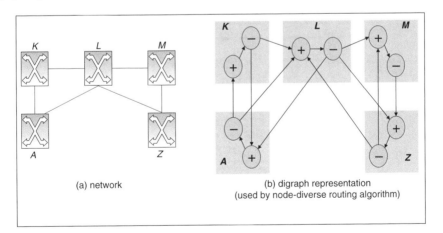

(a) network   (b) digraph representation
(used by node-diverse routing algorithm)

Figure 4.2: Network and equivalent graph representation. (From [107], Figure 6. Reproduced by permission of © 2003 The International Society for Optical Engineering.)

---

[3]The cost of the optical line can be defined as a function of a number of parameters, such as distance, load, etc.

4.3.2   Shared Backup Path-Protected Lightpaths

With shared backup path protection, backup paths are predefined, but the cross-connections along these backup paths are not created until a failure occurs. During normal operation the spare optical channels reserved for protection are not used. Since the capacity is only *soft-reserved*, the same optical channel can be shared to protect multiple lightpaths, as long as those multiple lightpaths are not susceptible to a common failure that would cause the backup channels to become exhausted without all failed lightpaths being recovered.

4.3.2.1   Policies for Backup Channel Assignment

There are actually two policies to assign reserved backup channels to protection paths. A *failure-dependent* strategy assigns the reserved backup channels in real time after failure occurrence on a first-come, first-served basis depending on availability (category 3 in Table 3.1) [89, 93]. Note, however, that those channels are preidentified as backup channels, and that the surviving channels of failed working paths are not made available for recovery, as is otherwise the case with *stub release* in path protection [164]. The assumption is that a proper backup channel-provisioning scheme ensures that enough protection channels are reserved so that all lightpaths can be recovered in case of a single failure. A *failure-independent* strategy assigns the reserved backup channels at the time of lightpath provisioning prior to a failure occurrence (category 2 in Table 3.1). The advantage of the second approach is that during lightpath protection, the switches on the protection paths immediately and individually cross-connect to predetermined channels, based on the identifier of the lightpath being recovered. If the channels are not preassigned, adjacent switches on the backup path must agree on the channels to be used for recovering the lightpath before establishing a cross-connection. This approach requires a handshake protocol and inter-switch signaling, which may be time-consuming and inadequate if recovery speed is of the essence.

In some cases, the failure-dependent, or pooling, approach reserves less channels for an equivalent level of protection. Consider for instance Figure 4.3. The figure represents four primary lightpaths routed in an optical network and their respective backup paths outlined with dotted lines. In this

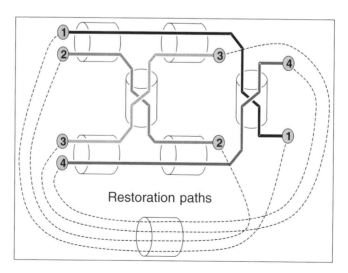

Figure 4.3: Pairwise full interference (optical switches are not pictured). (From [107], Figure 8. Reproduced by permission of © 2003 The International Society for Optical Engineering.)

example the optical switches are not pictured. The lightpaths are pairwise intersecting over six different links. The backup paths are all intersecting into one single fiber and may share reserved channels. A primary fiber cut in this example interrupts only two lightpaths, and thus two reserved channels are enough to recover the paths on a first-come first-serve basis. If the channels are predetermined in a failure-independent manner, four reserved channels, one for each backup path, are necessary to accommodate all existing combinations of lightpath pairs. However, experimental results indicate that in most situations both approaches are equivalent and the potential savings even if they exist, may not justify the additional complexity and processing time of the failure-dependent approach [51, 56, 89].

4.3.2.2 Protection against Single Link-SRG Failures

In the case of failure-independent channel assignment, there is a condition that two backup paths may share a reserved channel only if their respective primaries are link-SRG-disjoint, so that a failure does not interrupt both primary paths. If that happened, there would be contention for the reserved channel and only one of the two lightpaths would be successfully recovered.

Two lightpaths are said to be *mutually compatible*, if they are not affected by the same failure. Otherwise, they are *incompatible*. Figure 4.4(a) (for normal mode) and Figure 4.4(b) (for protection mode) picture an example of shared backup path protection, with the dashed lines representing reserved channels. Using the routing of Figure 4.4(a), demands AB and CD are compatible with respect to SRG failures and thus their backup paths share a single optical channel in link $S-T$, one less than in dedicated backup path protection. Upon failure as depicted in Figure 4.4(b), the ingress and egress nodes of the disconnected paths (X and Z in example) emit a request to the switches along the backup paths (S and T in example) to establish the cross-connections for that path. Once the cross-connections are established, each ingress and egress node switches the connection to the new path. This architecture requires fewer resources than in dedicated backup path protection. However, the protection involves more processing to signal and establish the cross-connections along the backup path. There is thus an evident trade-off between capacity utilization and recovery time.

Given an optical network with default SRGs, and a set of already established shared backup path-protected lightpaths, the problem of finding a feasible primary and shared backup path for a new lightpath request is NP-complete. The goal in such an approach would be to minimize the total cost of *new* optical lines used in the primary and backup paths. This essentially maximizes the sharing on the

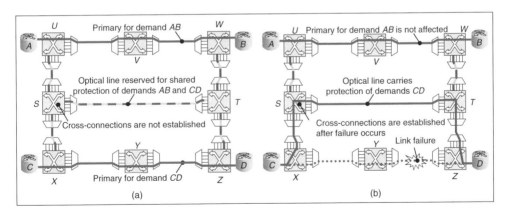

Figure 4.4: Shared backup path protection: (a) Network connections before a failure occurs (b) Network connections after a failure occurs. (From [107], Figure 7. Reproduced by permission of © 2003 The International Society for Optical Engineering.)

backup path, subject to latency constraints on primary/backup paths as well as protection switching time constraints. The proof of the NP-completeness of this problem is shown in the appendix of Chapter 5. A more detailed analysis of shared backup path protection in networks with SRGs is presented in Chapter 6.

4.3.2.3   Protection against Single Node and Link-SRG Failures

With shared backup path protection, node diversity between primary and backup paths does not guarantee full protection against node failures. Consider the example of Figure 4.5. The figure illustrates two primary bidirectional lightpaths (solid lines) and their corresponding backup paths (dotted lines) for demands (d, h) and (b, f). In this example, SRGs consist exclusively of all the optical channels on a link. The primary paths intersect on node i, while both secondary paths traverse links $e-f$ and $e-d$. The primary paths are SRG-disjoint, and so according to the sharing rules, the secondary paths may share reserved channels in their mutual links. If node i fails, reserved channels are allotted on a first-come first-serve basis and one demand is lost. This is an acceptable outcome if there is no commitment to recover either demand after a node failure. Otherwise the secondary paths cannot share channels if protection is also required in the event of a node failure for both demands. If it is desired that one demand only, say (h, d), survives a node failure, then in theory the backup paths may share reserved channels. However, in practice if node i fails, contention occurs for the reserved channels during the recovery process, and a first-come first-serve policy does not enforce systematic protection of (h, d), the demand that must be recovered. This problem can be averted by assigning a higher priority to (h, d), and if necessary preempting (b, f) in the recovery phase. Preemptive, priority-centric protocols entail processing overheads thereby increasing the recovery switching time, and are not elaborated upon in this chapter. If the recovery protocols do not allow preemption (or priorities), the routing algorithms must implement a more conservative redefinition of the sharing condition in order to avoid these pitfalls. As a result of these enhanced sharing conditions, the routing algorithm can guarantee the appropriate protection level for a lightpath.

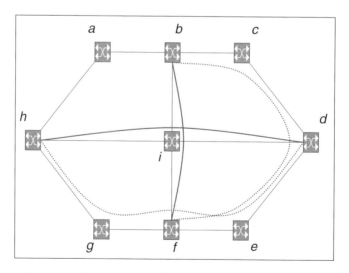

Figure 4.5: Shared backup path protection, with node disjoint paths but no protection against node failure. (From [107], Figure 9. Reproduced by permission of © 2003 The International Society for Optical Engineering.)

The provisioning of shared backup path-protected connections that guarantees protection against node failures entails more conservative rules of sharing. Assuming that

1. p_i and p_j denote two primary lightpaths,

2. S_i and S_j denote the SRG sets traversed by lightpaths p_i and p_j respectively,

3. N_i and N_j denote the node sets traversed by lightpaths p_i and p_j respectively,

4. $M_i \subseteq N_i$ and $M_j \subseteq N_j$ denote critical node sets (these are nodes for which protection is required in case of failure – typically all nodes along the path, except originating and terminating nodes), and

5. Primaries p_i and p_j must be recovered if a single failure occurs in $\{S_i, M_i\}$ and $\{S_j, M_j\}$ respectively.

Then p_i and p_j are *compatible* and their secondary paths may share channels if

- Rule (A): $S_i \cap S_j = \phi$, and

- Rule (B): $M_i \cap N_j = \phi$ and $M_j \cap N_i = \phi$.

Otherwise, p_i and p_j are said to be *incompatible (conflicting)*. Rules (A) and (B) express the condition that to be compatible, two primaries must be SRG and critical node-disjoint, that is they must not be affected by the same failure for which recovery is required.

Note that the current tenet in mesh networks is that path protection against single link failures protects against node failures for nodes of degree up to and including 3. This is true for dedicated backup path protection, but not for shared backup path protection, as demonstrated in the example of Figure 4.6. The example depicts two primary bidirectional lightpaths for demands (a, e) and (f, c). Both their respective backup protection paths traverse links $d-e$ and $c-d$. Node f is the source of (f, c) and an intermediate node of (a, e). Even though node f has degree 3, the sole provisioning of shared backup path protection for link failures would not protect the lightpaths against a failure of this node, since node c may initiate a request to reserve protection channels on edges $c-d$ and $d-e$,

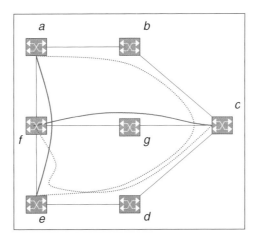

Figure 4.6: Example of node failures for nodes of degree up to and including 3. (From [107], Figure 10. Reproduced by permission of © 2003 The International Society for Optical Engineering.)

and thereby prevent the recovery of (a, e). If such level of protection is required, the rules of sharing stipulated above must be enforced with node f in the critical node list of demand (a, e).

Examples of algorithms for computing the routes of primary and backup paths when both single-node and SRG failures are taken into consideration are discussed in Chapter 6. The reader should also note that the preceding discussion only considered single failure scenarios. Even though multiple failure scenarios are not considered in this book except in Chapter 12 (in the context of service availability), there are numerous works in the literature that deal with this area of research. The reader is referred to [78, 96] for additional information on this subject.

4.3.3   Preemptible Lightpaths

Preemptible lightpaths are connections that use shared channels also assigned as part of the backup path of an SBPP connection. Preemptible lightpaths carry traffic until a failure forces an SBPP lightpath to establish and switch to its backup path. The preemptible lightpath gets preempted as some of the shared channels are taken away. Preemptible lightpaths can be established in two different ways. They can be restricted to only using already shared channels, or they can be allowed to use available channels which then become shared channels that can be used when establishing backup paths for SBPP lightpaths. In addition, preemptible lightpaths can automatically resume service once shared backup channels are reinstated after an SBPP lightpath reverts to its primary path.

Studies of mesh networks using span protection have shown that even surprisingly small fractions of preemptible lightpaths, mixed with protected, best-effort[4] and unprotected lightpaths, can be exploited in multi-Quality of Protection (QoP) designs to greatly reduce or even eliminate entirely conventional spare (unused standby) capacity in a survivable mesh network (see Section 5.9.3 in [139], [141]).

4.3.4   Diverse Unprotected Lightpaths with Dual-Homing

Unprotected lightpaths are different from preemptible lightpaths as their individual channels cannot be taken away from them. While an unprotected lightpath cannot individually protect against single failures, unprotected lightpaths can be combined to provide two diverse paths across the network in conjunction with dual-homing of the client onto two diverse access and egress nodes as shown in Figure 4.7. In such a recovery scheme, the protection switching that selects the best of the two incoming signals is performed by the customer equipment. Such a service can be modified by having one of the two diverse lightpaths as a preemptible lightpath, still offering a high level of availability at a lower cost.

Another variation would have the second path provisioned as a shared backup path, with the originating and terminating nodes of the backup path different from the originating and terminating nodes on the primary path as shown in Figure 4.8. With this protection scheme, shared backup channels would also be provisioned between the respective originating and terminating nodes for the primary path and the backup path so that the traffic on the primary path could be rerouted to the backup path. This protection scheme would be resilient against a combination of failures on the access side of the network (e.g., access toward the node that originates the backup path) and on a network link or node on the primary path. This protection scheme requires additional signaling capabilities so that the originating and terminating nodes can communicate and initiate the use of the shared backup channels between themselves.

---

[4]Best-effort lightpaths would be recovered through reprovisioning using spare capacity not assigned as backup capacity for protected lightpaths.

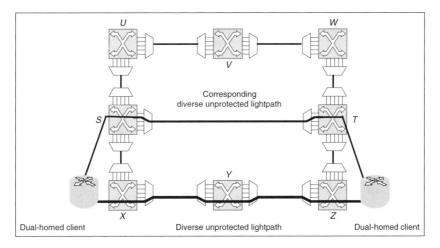

Figure 4.7: Example of two diverse unprotected lightpaths with dual-homing protection.

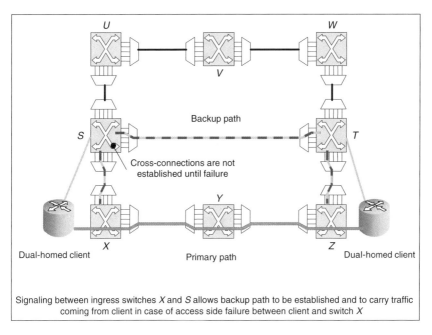

Figure 4.8: Example of primary and shared backup path-protected lightpath with dual-homing protection.

4.3.5   Multiple Simultaneous Backup Path-Protected Lightpaths

It is expected that connections with different protection schemes will coexist within the same mesh network. Service providers will be able to (1) change the routing of the backup path of a connection under a given protection scheme, and (2) change the type of backup path of a connection to migrate the connection to a different protection scheme. Such capabilities would apply across any type of survivability scenario (dedicated or shared backup path protection, etc.) and would allow the following activities to be accomplished by a network operator:

- Reoptimized the network by changing backup paths in such a way that at least one backup path is active all the time for a given primary path. This is required as the routing of primary and backup paths becomes non-globally optimal, due to network churn. A network operator will want to reoptimize the routing of the backup path[5] upon certain events (changes in the network, addition/removal of bandwidth, etc.), or on a periodic basis. If only one backup path can be defined at any time, the network operator needs to move the connection into a state where no backup path is available, before the new backup path can be defined. This leaves the connection unprotected for some finite amount of time, something that may not be desirable but tolerable. The reader is referred to Chapter 10 for an extensive analysis of the network reoptimization problem.

- Transition a connection from one protection type to another (e.g., from SBPP to DBPP) in such a way that at least one backup path is active all the time for a given primary path. Again, one approach would be to transition the connection through a state where no backup path is defined, thereby leaving the connection temporarily unprotected. This could be problematic because, if the provisioning of backup paths is done by a Network Management System (NMS), there is always the possibility that the NMS communication with the network could fail, leaving the connection in an unprotected state.

- Offer a service where more than one backup path is active at any one time. This would make the service more robust against multiple failures. This would also allow changing the set of backup paths while being able to maintain the k-recoverability of the connection.

In general, M dedicated backup paths and N shared backup paths could be defined at any one time. The resulting backup scheme can be described as (M, N) Multiple Backup Path Protection $((M, N)$ MBPP). Upon failures affecting the primary path and possibly some backup paths, the signaling would consider all the backup paths to recover the connection. For example, if there was no dedicated backup path left available, the signaling could initiate recovery on all the shared backup paths in parallel and select the first one that gets established for actual failure recovery. Figure 4.9 shows the case of a connection with a primary path (solid line), a dedicated backup path (dashed line), and a shared backup path (dotted line). In this case, $M = 1$ and $N = 1$. If the second backup path (dotted line) was changed to a dedicated backup path, this scenario would be qualified by $M = 2$ and $N = 0$. The standard DBPP and SBPP schemes are defined by $(M, N) = (1, 0)$ and $(M, N) = (0, 1)$, respectively.

4.3.6   Relaxing the Protection Guarantees

In the sections above, two classes of backup path-protected lightpaths were described, based on the protection guarantee: (a) single link-SRG failure protected lightpaths, and (b) single node and link-SRG failure protected lightpaths. Upon a single failure, 100% of all lightpaths (with the protection guarantee) that are affected by the failure are recovered. While routing a backup path-protected lightpath, spare capacity is allocated and assigned as backup capacity in such a manner as to provide the appropriate guarantee.

Service providers may desire the flexibility to provision backup path-protected lightpaths that (a) may require weaker protection guarantees or (b) may violate the protection guarantees under certain conditions. An example of case (a) above is when the service provider wishes to protect a lightpath against the failure of only a subset of critical SRGs (and nodes) in the network. An example of case (b) is when (in order to reduce blocking of new provisioning requests and prolong network lifetime),

---

[5]The reoptimization of primary paths is more delicate as it is not hitless and could impact on the service.

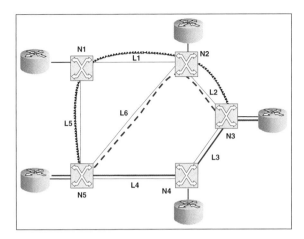

Figure 4.9: Example of multiple backup paths.

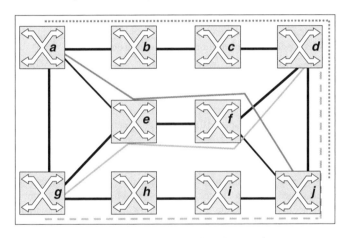

Figure 4.10: Network example with relaxed protection guarantees.

the service provider is willing to tolerate the loss of a certain fraction of lightpaths upon single failures of links and nodes. This section examines two ways in which protection guarantees may be relaxed from an algorithmic standpoint.

Consider the example in Figure 4.10. There are two shared backup path-protected lightpaths (a, j), and (g, d), each with a single-link failure protection guarantee. The primary (shown as solid line) and backup paths (shown as dashed line) for lightpath (a, j) are (a, e, f, j) and (a, b, c, d, j) respectively. The primary (shown as solid line) and backup paths (shown as dashed line) for lightpath (g, d) are (g, e, f, d) and (g, h, i, j, d) respectively. Primary paths (a, j) and (g, d) share the link $\{e, f\}$, therefore, their backup paths must not share any channels. In particular, on link $\{d, j\}$ the backup paths must use different channels.

Consider now the following two ways in which the protection guarantees can be relaxed:

1. Suppose that the protection guarantee is relaxed as follows: lightpath (a, j) needs to survive the failure of only a subset of critical links $\{\{a, e\}, \{f, j\}\}$ on its primary path, and lightpath (g, d) needs to survive the failure of only a subset of critical links $\{\{g, e\}, \{f, d\}\}$ on its primary

path. In this case, since the primary paths for lightpaths (a, j), and (g, d) intersect on link $\{e, f\}$ which is not a critical link, the two lightpaths can share channels on their backup paths. Each lightpath has a critical set of nodes and links that may be a subset of elements on its primary path, and a single failure in the critical set must be survived. The routing algorithms presented in Chapter 6 can be easily enhanced to handle this form of relaxation of the protection guarantee.

2. Suppose that lightpath (g, d) with a single-SRG failure guarantee is already provisioned with primary path (g, e, f, d) and backup path (g, h, i, j, d). Lightpath request (a, j) arrives, requiring a single-SRG failure guarantee. For lightpath request (a, j) consider primary path (a, e, f, j) and backup path (a, b, c, d, j). In this case, primary paths for lightpaths (g, d) and (a, j) share link $\{e, f\}$, and cannot normally share a channel on their backup paths on link $\{d, j\}$. However, if they are allowed to share a channel on their backup paths, thereby oversubscribing that channel, the protection guarantees of each lightpath are *degraded*. In particular, one or both lightpaths will not survive the failure of link $\{e, f\}$, since their backup paths will contend for the shared backup channel on link $\{d, j\}$. Service providers may oversubscribe shareable channels and allow protection guarantees to degrade for a variety of reasons:

   - Suppose that in the above example, lightpath (a, j) is a transient lightpath (held for a short period of time). The service provider may be willing to tolerate degraded protection guarantees for the single-link failures in the transient period.

   - Suppose that the network is overloaded. The network operator can add capacity to reduce the risk of blocking new provisioning requests. Alternatively, the network operator can prolong the lifetime of the network and reduce such blocking by degrading the protection guarantees for existing lightpaths. In this case, the network operator can recover capacity by *reoptimizing* (see Chapter 10) all the (degradable) backup paths to meet a degraded protection guarantee.

In the second case above, the network operator trades off protection capacity against the degradation of the protection guarantee (due to oversubscription of protection capacity). In general, the lightpaths may be partitioned into two classes, the degradable lightpaths and the nondegradable lightpaths. Degradable lightpaths may tolerate a degradation of their protection guarantee either because they are transient, or for other reasons. Nondegradable lightpaths may not tolerate a degraded protection guarantee.

There are several ways in which protection guarantees can be relaxed in a controlled fashion. For example, each lightpath may carry an attribute that indicates if its protection guarantee can be degraded. Or a network operator could define a network-wide protection guarantee which may be adjusted to require *that at least a specified percentage of the failed lightpaths are recovered upon any single failure.*

A possible algorithm for routing primary and backup paths with relaxed protection guarantees would simply set a higher weight or cost on links with a shared channel that violates the degraded protection guarantees (compared to those that do not violate the guarantees).

4.3.7 Impact of Multi-Port Card Diversity Constraints

Multi-port card or circuit-pack diversity constraint was introduced in the previous chapter as a new type of SRG, which can be referred to as a Shared Equipment Risk Group (SERG). The remainder of this section presents circuit-pack diversity routing algorithms, assuming a shared backup path-protected architecture. With shared backup path protection, two circuits must be link and circuit-pack diverse to be able to share a backup channel. The impact of derating on the cost-efficiency of the network is then measured under various demand forecast scenarios.

4.3.7.1 Algorithms for Multi-Port Card Diversity

The routing of a protected circuit arises principally under two contexts. In online routing, the network physical topology, the circuit-pack configuration, and the available capacity are prescribed inputs. Under this context, circuits are routed sequentially in the prescribed order, and strictly in accordance with the constraints imposed by the inputs. In particular, having a given circuit-pack configuration allows the algorithms to express and solve the circuit-pack diversity constraint in terms of node diversity constraints. In the context of an offline design, some of the constraints are relaxed, such as the configuration of the circuit packs, or the sizing of the network capacity, both of which can be optimized in the process in order to obtain the most cost-efficient solution. The offline design problem is investigated here. The routing and circuit-pack assignment problems are solved by first finding a cost-efficient routing of the demands, assuming shared backup path protection, and without

Table 4.1: Port to circuit-pack allocation algorithm

Definitions:

- For each demand (A, Z) of demand set T, let $r(A, Z) = (\{p, q\}, length)$ be the primary and backup path pair of demand (A, Z), and their *combined length* $= length(p) + length(q)$.

- Initially $r(A, Z) = (\{\emptyset, \emptyset\}, 0)$.

- Let R denote the set $\{r(A, Z)\}$ for all (A, Z) of T.

Algorithm:

1. R is given.

2. Set *min-total-length* $=$ infinity.

3. For each optical switch S of the network:

   - Let Q define the set of conflicting port pairs on S. Pairs of ports in that set cannot be collocated on the same circuit pack. Set $Q \leftarrow \{\}$.

   - For each pair of ports p_1 and p_2 of S: Using routing configuration R, if p_1 and p_2 are used by the primary and backup of the same demand (A, Z), or if p_1 and p_2 are used for the primaries of two demands that share a same backup channel, then set $Q \leftarrow Q + \{p_1, p_2\}$.

   - Rank ports of S, such that ports appearing the least in Q have lowest rank.

   - Sort ports according to increasing rank order; for each port p_i in that order:

     * If p_i can be located to an existing circuit pack C on S, then locate it to C.

     * Otherwise, if there is no circuit-pack on S, or if either all the ports of the circuit packs are used or the circuit packs already contain ports that are in conflict with p_i, then create new circuit pack on S and assign port p_i to it.

---

considering the circuit-pack diversity constraints. The ports required by the resulting routes are then grouped into circuit packs using a set-cover heuristic, so that the number of circuit packs is minimized while at the same time the circuit-pack diversity required for the shared backup path protection is respected. This algorithm is sketched in Table 4.1.

The input to the algorithm is a routing configuration and channel assignment that accommodates the prescribed demand set with protection against single SRG failures. Using this routing configuration, for each switch, all pairs of ports are identified that are in conflict with the diversity and sharing requirements. According to these requirements, on the same circuit pack: (1) two ports that are respectively traversed by the primary and the backup of a demand (diversity constraint) cannot be collocated, and (2) two ports that are traversed by the primaries of two demands sharing the same backup channel (sharing constraint) also cannot be collocated. These constraints are modeled by way of a conflict graph for each optical switch. The vertices of the conflict graph represent the ports of the optical switch, and the links indicate pairs of conflicting ports. The inverse of the conflict graph is the association graph, in which links indicate pairs of ports that can be collocated on the same circuit pack. Theoretically, if the algorithm enumerates all the cliques $\{K_i\}$ of the association graph, eliminates the cliques that exceed the size of a circuit pack, and groups the vertices (i.e. ports) of the remaining cliques into sets $\{C_i\}$, then finding the optimum allocation of ports to circuit packs requires finding the minimum set number that covers all the ports. The reader should note that a routing and channel assignment algorithm designed to protect against node failures always results in a fully connected association graph, since under these diversity and sharing constraints, ports on the same switch cannot be in conflict with each other [107]. The routing and channel assignment, the clique enumeration, and the set-cover problems are known to be NP-complete [128]; and all of them must be solved simultaneously if the optimal solution is sought. In practice these steps are solved independently, and a greedy algorithm described in Table 4.1 is used for the port to circuit-pack allocation after computing the routing and channel assignment problems.

4.4   Experiments and Capacity Performance Results

4.4.1   Performance Results for Path-Based Protection Techniques

All simulation experiments were run on two networks. N_{17} is a 17-node, 24-edge network that has a degree distribution of $(8, 6, 1, 2)$ nodes with respective degrees $(2, 3, 4, 5)$. N_{100} is a 100-node, 137-edge network that has a degree distribution of $(50, 28, 20, 2)$ nodes with respective degrees $(2, 3, 4, 5)$. These are realistic architectures representative of existing topologies. It is assumed that these architectures have infinite link capacity, and SRGs comprise exclusively all the optical channels in individual links. Five network robustness scenarios are considered: no protection; dedicated backup path protection against single link failures; dedicated backup path protection against single link or node failures; shared backup path protection against link failure; and shared backup path protection against single link or node failures. In network N_{17}, demand is uniform, and consists of two bidirectional lightpaths between every pair of nodes. This amounts to 272 lightpaths. In network N_{100}, 3278 node-pairs out of 4950 possible node-pairs are connected using one bidirectional lightpath per connection. Requests for lightpaths arrive one at a time (online routing) in a finite sequence and in an order that is arbitrary but common to each scenario to ensure a fair comparison. Figures of merit are capacity requirements separated into their primary and protection parts, and expressed in units of bidirectional OC-48 channels.

Results are presented in Figures 4.11 and 4.12. The quantities shown on the charts are averages over a series of 10 experiments using various demand arrival orders. These results indicate that link-disjoint and node-disjoint approaches[6] consume approximately the same total amount of capacity.

---

[6]Which insure guaranteed recovery against single link and node-and-link failures respectively.

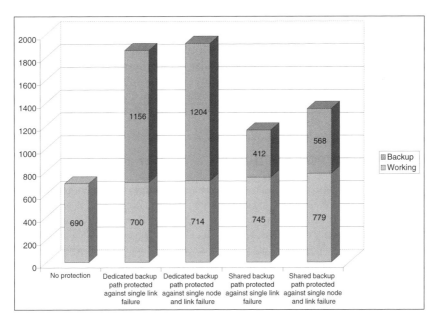

Figure 4.11: Comparison of capacity usage for different path protection architectures (17-node network). (From [107], Figure 11. Reproduced by permission of © 2003 The International Society for Optical Engineering.)

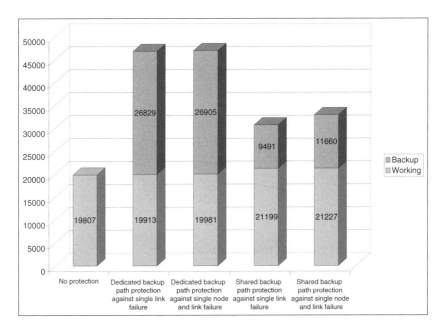

Figure 4.12: Comparison of capacity usage for different path protection architectures (100-node network). (From [107], Figure 12. Reproduced by permission of © 2003 The International Society for Optical Engineering.)

This is expected since dedicated backup path protection against link failures protects against node failures as well for nodes up to and including degree 3, and these nodes constitute a majority of all the nodes in the networks. This property does not apply to shared backup path protection, and to lightpaths that must be protected against node failures, even for nodes of degree 3. Thus, the provisioning of shared backup path protection to recover from node failures requires (in comparison to the dedicated schemes) more resources than to recover from link failures. The relative difference, however, is negligible considering the benefits of protecting the network against node failures.

4.4.2   Experiments with Multi-Port Card Diversity

A 45-node topology, with various demand loads, is used for experiments on multi-port card diversity. For each scenario circuit-pack sizes of 8 and 16 ports are assumed. Each case is then solved once with link but without circuit-pack diversity constraints (no Circuit-Pack Diversity Constraint (CPDC)), and once with circuit-pack diversity constraints as well (with CPDC). The two solutions are then compared in order to determine the combined effect of circuit-pack sizes and circuit-pack diversity on the network cost. The results of these experiments are summarized below in Figures 4.13 and 4.14. Figure 4.13 shows the number of circuit packs that are required to carry the demand as the load on the network increases. The load is expressed in units of the total number of ports required to accommodate the demand set. As expected, the number of circuit packs increases linearly with the demand load. The impact of the circuit-pack diversity constraint can also be significant for the larger circuit-pack size when the demand load is relatively small, but rapidly becomes negligible as the demand increases or the size of the circuit pack is reduced. Figure 4.14 shows the average circuit-pack utilization as a function of the total number of ports used by the demand. For instance, in the network under consideration, under heavy demand load, if each circuit pack can accommodate 16 ports, then only 80% of the circuit-pack capacity is used on average.

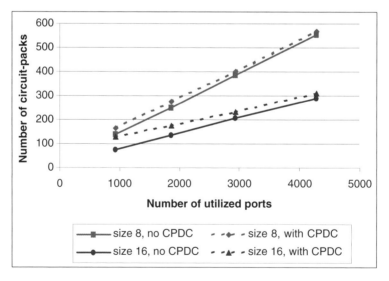

Figure 4.13: Number of circuit-pack modules required under various demand loads, circuit-pack sizes, and diversity constraints. (From [53], Figure 5. Reproduced by permission of © 2006 The Optical Society of America.)

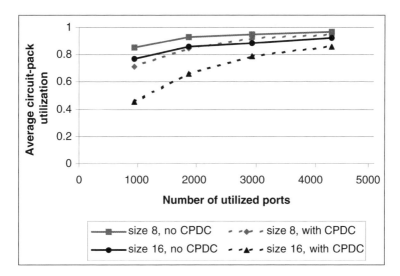

Figure 4.14: Average circuit-pack utilization under various demand loads, circuit-pack sizes, and diversity constraints. (From [53], Figure 6. Reproduced by permission of © 2006 The Optical Society of America.)

4.5 Recovery Time Analysis

Both simulation and analysis were used to calculate the recovery or protection switching time performance of shared backup path-protected networks. For the simulation approach, an optical mesh network modeling and simulation tool was used that models the protection architecture and protocols, including delays and other processing in the switch hardware, messaging sequences, etc. [46]. Two networks are used for the simulation studies. A 17-node network is illustrated in Figure 4.15 and a 50-node network shown in Figure 4.16. For the first network, the average node degree is 3.1 and there are 224 OC-48 lightpaths (with randomly selected end-nodes) routed using link-disjoint shared backup path protection. For the second network, the average node degree is 3.44 and there are 910 OC-48 lightpaths with randomly selected end-nodes routed using link-disjoint shared backup path protection. The reader should note that these figures show the SRG dependencies in the networks (via cylindrical-shaped icons on the links).

To test the performance of the shared backup path protection scheme in terms of protection switching time, the simulation studies for the 17-node network involved failing single conduits, which in turn resulted in the simultaneous failure of the multiple working paths traversing these conduits. In that case, the average protection time can be calculated from the time it takes for all the affected lightpaths to be recovered. The maximum protection switching time in this case will be defined as the time it takes to recover the last affected lightpath. Figures 4.17 and 4.18 show the results for maximum and average protection time respectively for the 17-node network. Figure 4.17 illustrates that even though the numbers of lightpaths affected by the three failure instances are relatively close, protection switching times vary and they do not necessarily increase proportionally to the number of affected lightpaths. This is mainly due to the fact that other parameters, such as the number of failed lightpaths that are processed by a single node, topology, traffic load, distribution of lightpath endpoints, etc., come into play for the calculation of the protection time. Figure 4.18 shows that, in general, the protection time increases with the number of failed lightpaths, even though there could be variations for a relatively similar number of failed lightpaths as shown in Figure 4.17 [16].

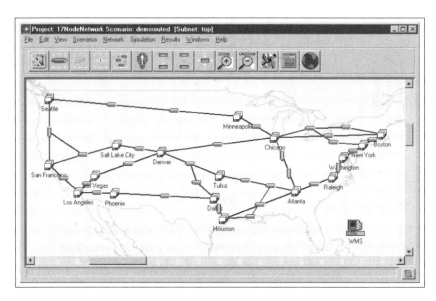

Figure 4.15: A 17-node network used for simulating protection time performance in shared backup path protected networks. (From [16], Figure 5. Reproduced by permission of © 2002 The Optical Society of America.)

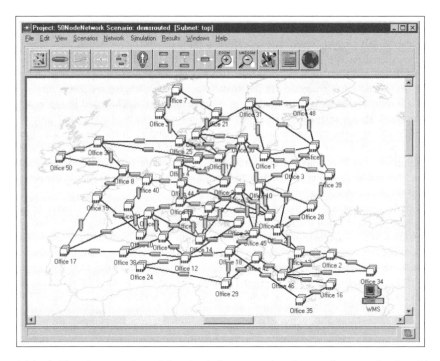

Figure 4.16: A 50-node network used for simulating protection time performance in shared backup path-protected networks. (From [16], Figure 8. Reproduced by permission of © 2002 The Optical Society of America.)

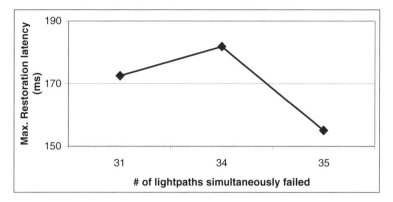

Figure 4.17: Maximum protection time experiments for a 17-node network. (From [16], Figure 6. Reproduced by permission of © 2002 The Optical Society of America.)

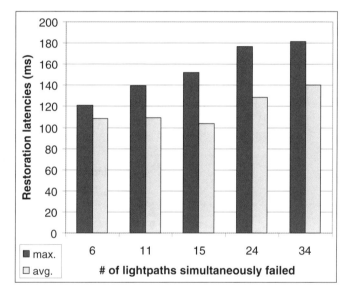

Figure 4.18: Average and maximum protection time experiments for a 17-node network. (From [16], Figure 7. Reproduced by permission of © 2002 The Optical Society of America.)

The simulation studies for the 50-node network involved failing the conduits that carried the maximum number of lightpaths, so as to determine the worst case protection switching times. Figure 4.19 illustrates these results. From this figure it is apparent that the protection switching time for this network configuration is capped at approximately 200 ms [16].

An approximate analytical approach was also developed to validate the simulation results [16]. The assumptions used in this approach are that the traffic demand is uniform and that there is a linear dependency between the protection time and number of lightpaths recovered. Given that the average number of lightpaths per link is k, the average protection switching time can then be approximated using the worst case assumption that all k lightpaths that are affected by a failure terminate at the same two switches and that a failure occurs in the middle link (SRG) of the primary path (in terms

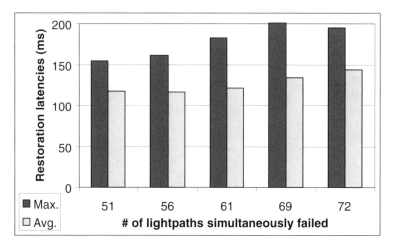

Figure 4.19: Average and maximum protection time experiments for a 50-node network. (From [16], Figure 9. Reproduced by permission of © 2002 The Optical Society of America.)

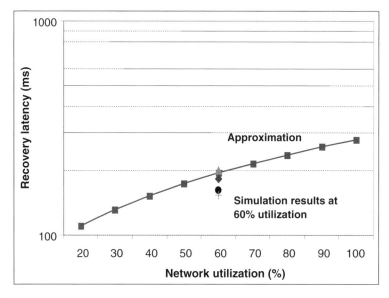

Figure 4.20: Analytical vs simulation results for a 50-node network. (From [16], Figure 10. Reproduced by permission of © 2002 The Optical Society of America.)

of number of hops). The average protection time can then be defined as

$$T_f = T_1 + (k-1)L \tag{4.1}$$

where T_f is the time when all lightpaths have been recovered, T_1 is the time required for the recovery of the first lightpath and L is a parameter that represents the slope of the linear tail of protection time versus number of lightpaths recovered. To calculate T_1, a single lightpath failure is assumed [16].

Analytical results with the assumptions outlined above are shown in Figure 4.20 and are compared with the simulation results of the 50-node network. It is shown that at 60% utilization, the analytical approximation yields results of the same order of magnitude as the simulation results for five single failure events affecting the most number of lightpaths. Further details and discussions on recovery time performance in path-protected mesh networks can be found in Section 11.8.

4.6 Recovery Time and Capacity Trade-Offs

Table 4.2 summarizes the protection architectures presented earlier in this chapter, the performance results for the different architectures and their complexity for experiments utilizing centralized routing. For each protection architecture and SRG type, the complexity of the lightpath provisioning operation (polynomial or NP-complete), the speed of protection as well as the cost of the service expressed in amount of resources used by the protection mechanism (as percentage of working capacity) are indicated. Note that all the percentages mentioned in this table are based on a particular set of experiments.

From the discussion above, it is clear that provisioning of backup paths in shared backup path protection architectures sometimes requires longer paths that consist exclusively of existing shared backup channels, rather than shorter paths where spare channels must be turned into shared backup channels. Figures 4.21, 4.22 and 4.23 are examples that clearly demonstrate this case. In these figures working connections (D, F), (D, E) and (E, F) and their respective backup paths (D, A, B, C, F), (D, G, H, E) and (E, H, I, F) are already provisioned, and a new primary path (A, B) arrives at the network. Clearly, if the backup path (A, D, G, H, I, F, C, B) is used – (Figure 4.21), no additional protection resources are required. However, uncontrolled sharing may result in arbitrarily long backup paths, at the expense of recovery time. At the other extreme, if the backup path (A, D, E, B) is used (Figure 4.22), it will result in the shortest possible backup path (faster recovery time) but will restrict sharing opportunities, which in turn means that it will require more ports, and thus cost more. A middle-of-the-road solution of the backup path $((A, D, G, H, E, B)$–Figure 4.23) will result in a backup path of average length and with some sharing of the protection resources.

Several additional experiments were performed to evaluate the effect of additional hops allowed on backup path over the shortest-hop path. Figure 4.24 shows the effect on the protection capacity required for a 17-node network of allowing additional hops in the backup path, over the shortest path alternative. Seven experiments were performed. In experiment indexed by number j the length of each protection path is constrained to be at most the shortest-hop path plus j hops. The plots indicate the amount of protection capacity as a percentage of the working capacity, which does not vary across the experiments. These results demonstrate that approximately 17% savings in protection

Table 4.2: Summary of different architectures and their complexity (centralized routing). (From [107], Table 2. Reproduced by permission of © 2003 The International Society for Optical Engineering.)

Protection architecture	SRG type	Complexity of centralized routing	Protection resources (% of working capacity)	Speed of protection
SRG failure dedicated backup path protection	(a), (b)	Polynomial	100–170	Very fast
	(c)	NP-complete	> 100	Very fast
SRG and node failure dedicated backup path protection	(a), (b)	Polynomial	100–175	Very fast
	(c)	NP-complete	> 100	Very fast
SRG failure shared backup path protection	(a), (b), (c)	NP-complete	40–70	Fast
SRG and node failure shared backup path protection	(a), (b), (c)	NP-complete	40–80	Fast

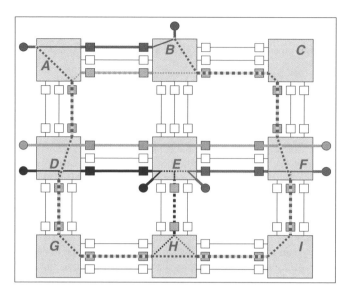

Figure 4.21: Cost vs latency with shared backup path protection – Example 1.

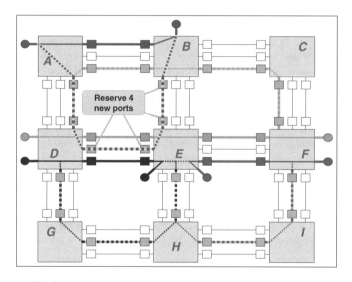

Figure 4.22: Cost vs latency with shared backup path protection – Example 2.

capacity requirements in this example, are possible by allowing longer protection paths. The graph also indicates that four additional hops over the shortest-hop paths are sufficient to gain most of the benefits. In these experiments, the average protection path hop distance increases 25% from four to five hops.

Various trade-offs between the backup path length and the protection capacity for shared backup path protection (by limiting the length of the backup path, or changing the cost of using shared channels) are reported in [187]. In these cases the questions that one should ask are: (a) What is the trade-off between shareability and protection time? and (b) How is this trade-off achieved? Chapter 7

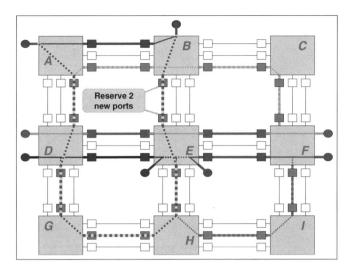

Figure 4.23: Cost vs latency with shared backup path protection – Example 3.

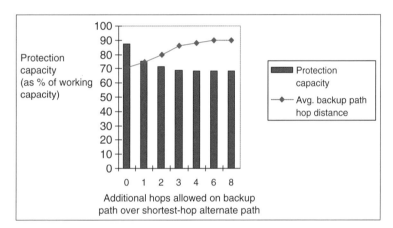

Figure 4.24: Effect of additional hops allowed on the backup path over the shortest path alternative on the protection capacity required. (From [187], Figure 12. Reproduced by permission of © 2002 Kluwer Academic Publishers.)

deals precisely with this issue and defines new metrics that are used to find *good* primary and backup paths (minimize the capacity used, while at the same time restrict the length of the backup path).

4.7 Conclusion

In this chapter we have focused on routing and protection in path-protected mesh networks, specifically DBPP, SBPP, and variations thereof. Focusing our attention on the OXC-based protection in the optical domain only, we reviewed and compared through simulation results a set of protection architectures. In order to conduct these comparisons we first described multiple scenarios based on architecture characteristics (e.g., Shared Risk Groups), the type of failures for which protection was required (e.g., node or link) and whether the provisioning of lightpaths was centralized or distributed. We then compared available protection mechanisms taking into consideration figures of merit such as protection capacity and speed of recovery to sieve out the most appropriate protection schemes for each failure scenario.

Experimental results show that the amount of resources used protecting node failures using the dedicated backup path protection mechanism is only marginally higher than dedicated backup path protection against SRG failures, and the difference may not justify the implementation of a dedicated protection mechanism that protects against SRG failures only. If minimizing the cost of the service is the main objective, then shared backup path protection, which allows sharing of protection channels among the backup paths, is more appropriate. Shared backup path protection also has the additional advantage over dedicated protection that it promotes preemptible services that use idle protection channels, giving way to higher-priority services when a failure occurs. However, unlike dedicated protection, shared backup path protection against node failures consumes substantially more resources than shared backup path protection against link failures. Thus, if shared backup path protection is utilized, there is a trade-off in terms of enhanced network resilience versus network capacity and subsequently network cost.

Other important aspects that have been put aside in this chapter, and are good candidate topics for future work, are the constraint of round-trip delays and simulation of protection switching times under various load conditions. For a fair comparison with other types of protection, it is important to limit the length of the backup protection paths in order to respect prescribed round-trip latencies and provide equivalent level of services on primary and backup paths. Reported experiments indicate that protection paths with unrestricted lengths may reserve an arbitrarily large number of channels, although on the average the length does not increase much compared to the shortest-hop alternative.

Chapter 5

Path Routing – Part 1: Complexity

5.1 Introduction

This chapter examines the algorithmic complexity of routing problems. Algorithms for routing in networks are a classical area of study spanning many decades. The emergence of broadband networks and optical networks have presented a new set of routing problems and the complexity of such problems will be the focus of this chapter. The routing problem in a circuit-switched or path-oriented mesh network is the following: given the network, and given a service to be provisioned in the network, find a route that optimizes certain metrics subject to the constraints of the network state, and of the service requirements. The route specifies the nodes, links and optionally the channels and the bandwidth slots of channels on which the service is provisioned. A service request is specified by a bandwidth requirement, and a protection requirement. Depending on the protection requirement, the routing problem involves finding a working and a protection route. This chapter examines the complexity of routing working and protection paths in mesh networks. The next chapter (Chapter 6) explores practical heuristics to solve the routing problems presented in this chapter.

5.2 Network Topology Abstraction

An algorithm for routing uses an abstraction of the network topology and the service to be routed. Such an abstraction is first presented in the following. A mesh network can be modeled as a directed graph $G(V, E)$ of a set of nodes V interconnected by a set of directed links E. Each link (u, v) from node u to node v has the following attributes:

- Metrics such as cost and delay. Such metrics may be defined to be independent of each other.

- A set of diversity attributes or Shared Risk Group (SRG) identifiers. SRGs are assigned identifiers that are unique network-wide.

- A set of channels. Each channel is a pool of bandwidth, and represents the following attributes:

    - Maximum bandwidth of the channel.

Path Routing in Mesh Optical Networks   Eric Bouillet, Georgios Ellinas,
Jean-François Labourdette, Ramu Ramamurthy   © 2007 John Wiley & Sons, Ltd

– Available bandwidth: bandwidth that is available for use. Available bandwidth may be represented either as an aggregate quantity or may provide a detailed mapping of bandwidth units that are available on the channel.

– Quality-of-Service (QoS) classes pertaining to the link. Channels on the link may be classified into different classes depending on user configuration or depending on how the classes are treated for buffering/scheduling.

– Link-layer protection attributes for the link. Link-layer protection schemes include standardized protection schemes such as 1+1 APS.

5.2.1 Service Definition

Chapter 4 (Section 4.2) examines in detail the metrics and constraints that define services. For the purposes of this chapter, a service may be modeled by the following attributes:

• Bandwidth required for the service.

• QoS attributes of bandwidth (type of channel to be used) on each link of the path.

• Protection attributes: whether the service needs to be protected upon the failure of an SRG and/or a node on the path. A protected service between node u and node v is represented by a pair of SRG-diverse (or node-diverse/node-and-SRG-diverse) paths between u and v. This attribute also indicates the type of recovery scheme to be utilized, dedicated or shared backup path protection. In DBPP, channels on the backup path are reserved for the service, whereas in SBPP, channels on the backup path may be shared with other services.

5.2.2 Operational Models: Online vs Offline Routing

The application dictates the nature of routing algorithms that are employed. There are the following two models for route computation:

• Online Routing: in this model, route computation for a service is performed in real time when the service is required to be provisioned. This scenario corresponds to a dynamic network setting where services are provisioned and deprovisioned frequently. The routing algorithm is expected to respond quickly to the provisioning request. In this framework, optimality of the route is less important than finding a feasible route quickly.

• Offline Routing: in this model, route computation for services is performed separately from provisioning. During network design and planning activities, services may be routed although they may not be provisioned. In this framework, optimal routes for services are desired. The routing algorithm is not time-constrained, and usually routes multiple services simultaneously.

This chapter examines online algorithms for route computation, and an Integer Linear Programming (ILP) framework for offline route computation.

5.3 Shortest-Path Routing

Shortest-path routing is at the heart of the routing algorithms presented in this and the next chapter. The objective for shortest-path routing is to minimize the cost of the links used on the path. The shortest-path routing problem is solved by two classical algorithms: (a) the Bellman-Ford algorithm [34, 122] and (b) Dijkstra's algorithm [92]. While the Bellman-Ford algorithm allows for negative edge weights, Dijkstra's algorithm requires nonnegative edge weights. However, Dijkstra's

Table 5.1: Dijkstra's algorithm

Definitions:

- $G(V, E)$: weighted directed graph, with set of vertices V and set of directed edges E

- $w(u, v)$: cost of directed edge from node u to node v (costs are nonnegative). Links that do not satisfy constraints on the shortest path are removed from the graph

- s: the source node

- B: is a heap data-structure

- S: set of nodes to which the shortest path has already been found

- $d(u)$: cost to node u from node s, equal to the sum of the edge costs on path from s to u

- $p(u)$: node previous to u on path from s to u

Algorithm:

- $S = \phi$

- $d(s) = 0$, $d(u) = \infty (\forall u \neq s)$, $p(u) = 0 (\forall u)$

- insert all nodes into B

- while B is not empty

  - let u be the minimum cost vertex in B
    * if $d(u) = \infty$ then exit
  - $S = S \cup \{u\}$, $B = B - \{u\}$
  - for each vertex v adjacent to u do
    * if $d(v) > d(u) + w(u, v)$ then $d(v) = d(u) + w(u, v)$, $p(v) = u$

---

algorithm scales better with the size of the network, and in most practical networks edge weights are nonnegative. Dijkstra's shortest-path routing algorithm is widely implemented in internet routing protocols such as Open Shortest Path First (OSPF) and Intermediate System–Intermediate System (IS-IS), and is used as the basic building block for various routing algorithms. Dijkstra's algorithm simultaneously finds the shortest-cost path from a given source node to all other nodes in the network.

5.3.1 Dijkstra's Algorithm

Dijkstra's algorithm is illustrated in Table 5.1. It maintains two sets of nodes: nodes reachable from the source node are in one set (denoted by S in the algorithm), and nodes not yet known to be reachable are in another set (denoted by B in the algorithm). In each step of the algorithm, a node is selected from B and added to set S. The output of Dijkstra's algorithm is the minimum cost path tree from a given node s to all other nodes reachable from s.

Table 5.2: Generalization of Dijkstra's algorithm to find K-shortest paths (paths may contain loops)

Definitions:

- $G(V, E)$: weighted directed graph, with set of vertices V and set of directed edges E

- $w(u, v)$: cost of directed edge from node u to node v (costs are non-negative). Links that do not satisfy constraints on the shortest path are removed from the graph

- s: the source node

- t: the destination node

- K: the number of shortest paths to find

- p_u: a path from s to u

- B is a heap data structure containing paths

- P: set of shortest paths from s to t

- $count_u$: number of shortest paths found to node u

Algorithm:

- P = empty, $count_u = 0, \forall u \in V$

- insert path $p_s = \{s\}$ into B with cost 0

- while B is not empty and $count_t < K$:

    - let p_u be the shortest cost path in B with cost C
    - $B = B - \{p_u\}$, $count_u = count_u + 1$
    - if $u = t$ then $P = P \cup p_u$
    - if $count_u \leq K$ then
        * for each vertex v adjacent to u:
            - let p_v be a new path with cost $C + w(u, v)$ formed by concatenating edge (u, v) to path p_u
            - insert p_v into B

When the heap is implemented as a Fibonacci heap [86], the running time of Dijkstra's algorithm is $O(V \log V + E)$.

5.3.2 Dijkstra's Algorithm Generalization to K-Shortest Paths

Finding more than one shortest path between a pair of nodes is often needed. Such paths can be used when there are additional constraints on the path that the shortest path does not meet. Dijkstra's algorithm can be easily extended [103, 280, 281, 282] to find more than one shortest path between a pair of nodes as illustrated in Table 5.2. However, the paths found by this algorithm will not necessarily be loopless.[1]

---
[1] Algorithms for K-shortest loopless paths are discussed in Section 6.1.5.

5.3.3 Shortest-Path Routing with Constraints

In practice, in addition to minimizing cost, several other nontrivial constraints are often imposed on a route. For example, the route must be of minimum cost while at the same time the total latency on the path must be bounded. Such a constraint is often imposed on latency-sensitive applications such as Voice over IP (VoIP). Dijkstra's algorithm cannot solve the minimum-cost routing problem with such constraints.

5.3.3.1 Constraint on Additive Link Metrics

Finding a feasible path which has a bounded cost and at the same time satisfies a constraint on an additive link metric such as delay, is an NP-complete problem [128, 312]. Jaffe [168] proposed the pseudo-polynomial approach illustrated in Table 5.3 to solve this problem.

Assume that the maximum delay on any link is an integral b. All loop-free routes in the graph will have a delay of at most Nb, where N is the number of nodes. The algorithm maintains a table

Table 5.3: Jaffe's algorithm for shortest-path with an additional constraint

---

Definitions:

- $G(V, E)$: directed graph, with set of nodes V and set of edges E. The number of nodes is N.

- $w(u, v)$: cost of edge from node u to node v.

- $d(u, v)$: delay on edge from node u to node v. The maximum delay on any edge is bounded by b. The maximum delay on any loop-free path is bounded by Nb.

- $l(u, v, k)$: cost of path to node v from node u with delay at most k, where k ranges from 1 to Nb.

- $changed(u, v, k)$: if the table entry $l(u, v, k)$ was changed in the previous iteration of the algorithm

Algorithm:

1. $changed(u, v, k) = $ false for all u, v, k.

2. Initialize $l(u, v, k)$ as follows for all u,v,k:

   - $l(u, v, k) = \infty$ if $(u, v) \notin E$ or $d(u, v) > k$

   - otherwise, $l(u, v, k) = w(u, v)$, $changed(u, v, k) = $ true

3. for $N - 1$ iterations do:

   - for all u,v,k such that $changed(u, v, k) = $ true do:

     – $changed(u, v, k) = $ false

     – for each node x such that $(x, u) \in E$ (where $d(x, u) = d$) do:

       * for $k' \geq k + d$ do: if $l(x, v, k') > l(u, v, k) + w(x, u)$ then $l(x, v, k') = l(u, v, k) + w(x, u)$, $changed(x, v, k') = $ true

for each node u, whose entry $l(u, v, k)$ contains the cost of the shortest path from u to v, whose delay is atmost k (k ranges from 1 to Nb). In each iteration, the table entries are updated by relaxing each edge in the graph as illustrated in Table 5.3. This algorithm finds the minimum-cost solution that meets the delay constraint in $O(N^6 b^2)$ time. The drawback of the above approach is that the running time is sensitive to the maximum value of the delay parameter. Chapter 6 examines another heuristic approach to this problem.

5.4 Diverse-Path Routing

This section examines diverse-path routing in mesh networks. Paths p_1 and p_2 are diversely routed if the set of SRGs on the links belonging to p_1 is disjoint from the set of SRGs on the links belonging to p_2. Paths p_1 and p_2 are node-diverse if p_1 and p_2 do not share any intermediate nodes. As noted in Chapter 4, SRG (or more specifically SRLG) diversity does not imply node diversity and node diversity does not imply SRG diversity. Two paths are node-and-SRG-diverse if they are both node-diverse and SRG-diverse.

In some network topologies or under certain network conditions (for example when the network is congested with many links close to capacity), there may not exist diverse primary and backup paths in the network between several node-pairs. In such cases, it may still be desirable to provision protected services by allowing the primary and backup paths to have SRGs in common, thus violating the diversity requirement.

5.4.1 SRG Types

Section 3.3 introduced the concept of a Shared Risk Group (SRG) and illustrated the three different types of SRGs as shown in Figure 5.1. The default SRG spans exactly one link and is illustrated as *Type a*. *Type b* SRGs (also called *fork* SRGs) span links each of which terminates on a common node. This is practically realized by a conduit which all the links originating at a node may traverse. *Type c* SRGs are the most general form of SRGs which may span any set of links. *Type a* SRGs are a subset of *Type b* SRGs and *Type b* SRGs are a subset of *Type c* SRGs. Such a classification of SRGs is useful in the discussion of diverse-path routing algorithms in the sections that follow.

5.4.2 Diverse-Path Routing with Default SRGs

When all SRGs are of *Type a*, the diverse routing problem is identical to the link-diverse routing problem in a network. Suurballe's algorithm [293] solves the general problem of finding K link (or

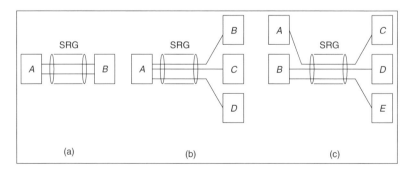

Figure 5.1: Different types of SRGs.

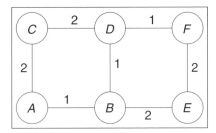

Figure 5.2: Trap topology for diverse routing. The shortest-cost path does not have a diverse path, even though diverse paths exist in the network.

node)-diverse paths in a network. Bhandari [40] is a good reference on diverse routing algorithms in mesh networks. The following discussion focuses on the more restricted problem of finding a pair of link (or node)-diverse paths in a network.

A diverse pair of routes in a directed graph can be determined as a two-step procedure by first finding the primary path, and then finding the diverse backup path after removing the edges (and nodes if node diversity is required) of the primary path. However, this approach can fail even when there is a feasible solution as illustrated in Figure 5.2. In this example, the shortest path $A-B-D-F$ (with cost 3) does not have a link-diverse path, whereas a longer path $A-B-E-F$ (with cost 5) has a diverse path $A-C-D-F$.

The edge connectivity of the network is the smallest set of edges that when removed, disconnects the network.[2] For edge-diverse routes to exist, the network must be two-edge connected (the removal of any one edge does not disconnect the network). Similarly, the vertex connectivity of the network is the smallest number of nodes which, when removed, disconnect the network. Diverse routing algorithms are closely related to network-flow techniques [14, 139, 238, 296]. Indeed, edge-diverse routing between a source node s and a destination node t can be considered equivalent to routing a flow of 2 units between the source and destination nodes in a network where each edge has capacity of 1 unit. Edge-diverse paths are then found as the paths constituted by the edges in the shortest path, and the augmenting path in the residual graph. Such an algorithm from [40] is illustrated in Table 5.4.

Figure 5.3 illustrates the application of the algorithm presented in Table 5.4 to the network of Figure 5.2. In Figure 5.3(a), $A-B-D-F$ is the shortest path from A to F. Figure 5.3(b) illustrates the shortest path tree from node A along with distances to each node from node A, Figure 5.3(c) shows each edge with an updated cost function, and the edges on the path from A to F reversed in direction, and Figure 5.3(d) shows the shortest path in the residual graph. Combining the edges in the two paths $\{A-B, B-D, D-F, A-C, C-D, D-B, B-E, E-F\}$, and removing the common edge $\{B-D\}$ between the two paths results in two diverse paths $A-B-E-F$ and $A-C-D-F$.

The algorithm described in Table 5.4 requires two iterations of the Dijkstra algorithm. To find diverse paths from a given node to all other nodes, a straightforward approach applies the above algorithm for each target node resulting in a running time of $O(V^2 \log V + VE)$. [292] presents techniques to speed up the above algorithm to have a running time of $O(E \log V)$.

When node-diverse paths are required, the following vertex splitting transformation illustrated in Figure 5.4 may be used. Each node v in the original graph G is split into two nodes v_1 and v_2 in the transformed graph G'. For each incoming edge (u, v), an edge (u_2, v_1) is added. For each outgoing edge (v, w) an edge (v_2, w_1) is added. The edge (v_1, v_2) of zero cost connects the two vertices. Edge-diverse paths in the transformed graph G' correspond to node-diverse paths in G.

---

[2]A network is disconnected when there exists a pair of nodes with no path between them.

Table 5.4:   Suurballe's algorithm for edge-diverse routing

Definitions:

- $G(V, E)$: weighted directed graph, with set of vertices V and set of directed edges E

- $w(u, v)$: cost of directed edge from node u to node v (costs are non-negative). Links that do not satisfy constraints on the shortest path are removed from the graph

- s: the source node

- t: the destination node

- $d(u)$: cost of path to node u from node s in the shortest path tree rooted at s

Algorithm:

1. find the shortest path tree, T, from node s. Let $P1$ be the shortest cost path from s to t

2. update the costs on each edge (u, v) as $w'(u, v) = w(u, v) + d(u) - d(v)$. All tree edges have cost 0, and non tree edges have non negative cost

3. create the residual graph by removing existing edges on path $P1$ directed towards s and reversing each edge on $P1$

4. find the shortest cost path $P2$ in the residual graph

5. the shortest pair of paths is determined by the edges in $P1$ and $P2$ after removing common edges between $P1$ and $P2$ ((u, v) in $P1$ and (v, u) in $P2$)

5.4.3   Diverse-Path Routing with Fork SRGs

In the mesh network topology, assume that for each SRG, all the links that contain that SRG have a node in common. Furthermore, each link adjacent to a node has at most one such SRG with that node in common. This situation corresponds to the practical case of conduits emerging from a node and the cables belonging to that conduit being routed to separate nodes. In this case, the routing graph can be transformed into an auxiliary graph that allows the use of diverse routing algorithms previously discussed.

For SRG-diverse routes the routing graph transformation is illustrated in Figure 5.5. Each node is split into one auxiliary node for each SRG that is adjacent to it. There is an auxiliary edge (with cost = 0) between each pair of such auxiliary nodes. Node-disjoint routes in the transformed routing graph correspond to SRG-disjoint routes in the original routing graph.

For node-diverse routes with *Type b* SRGs the routing graph transformation is illustrated in Figure 5.6. For each node, there is an auxiliary node for each SRG that is adjacent to it. The node and its auxiliary nodes are connected in a star fashion via auxiliary edges with cost 0. Node-disjoint routes in the transformed routing graph correspond to node-disjoint routes in the original routing graph.

5.4.4   Diverse-Path Routing with General SRGs

If general SRGs (of *Type c*) are present in the network, then the problem of finding a feasible SRG-diverse route in the network is NP-complete [107] as described in Section 5.8.1 (the appendix of

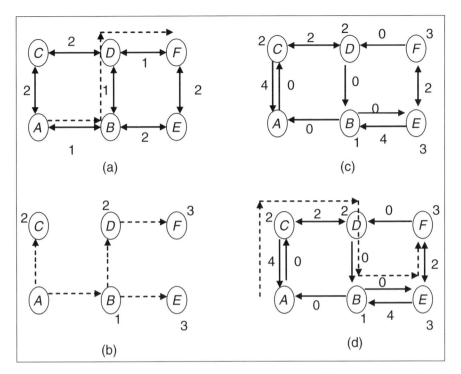

Figure 5.3: (a) Shortest path in the graph, (b) Shortest path tree with distances to each node, (c) Updated cost on each edge, with the edges on the shortest path reversed in direction, and (d) Shortest path on the residual graph.

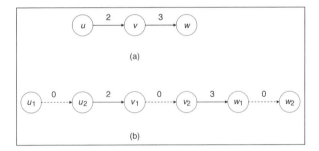

Figure 5.4: Vertex splitting transformation used to find node-diverse paths using the edge-diverse routing algorithm.

this chapter). The difficulty of finding feasible diverse routes arises due to the possibility of defining arbitrary SRGs. A solution is to restrict the definition of SRGs to those configurations that are algorithmically tractable. However, this restricts possible network configurations.

A heuristic for diverse routing in the general case resorts to an enumeration of primary paths as illustrated in Section 6.4. For each choice of a primary path, a backup path is found after removing the SRGs (and intermediate nodes if node-diverse routing is required) that are contained in the primary path.

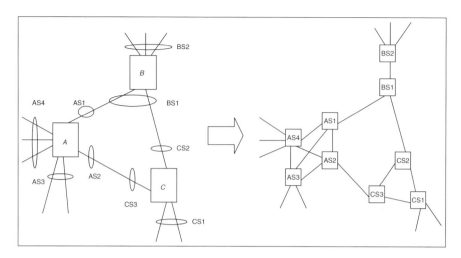

Figure 5.5: SRG-diverse routing transformation for *Type b* SRGs.

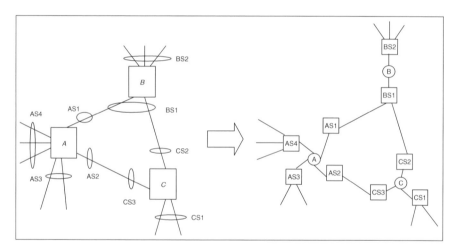

Figure 5.6: Node-diverse routing transformation for *Type b* SRGs.

5.5   Shared Backup Path Protection Routing

An SBPP service has a pair of diverse paths in the network, where one of the routes is a working route, and the other is a backup route. Each channel of the working route has dedicated capacity allocated to the service (and carries traffic under normal conditions). Each channel of the backup route also has capacity dedicated to this service; however, the capacity on the backup path can be shared with backup paths for other SBPP services. For any two SBPP services p_1 (working route = w_1, backup route = b_1) and p_2 (working route = w_2, backup route = b_2), b_1 and b_2 can share a channel if w_1 and w_2 satisfy the rules of sharing outlined in Section 4.3.2.3. A meaningful notion of the cost of an SBPP service p whose working route is w and backup route is b is defined as the sum of the costs of the channels in w and costs of the unshared channels in b. This definition sums up the costs of new capacity allocated to the service.

5.5.1 Protection Guarantees and Rules of Sharing

The rules of sharing determine the conditions under which a channel may be used for the backup path for an SBPP service. Intuitively, two services may share a channel on their backup path only if they do not share a failure mode, that will simultaneously bring them both down and cause them to compete for that channel. Section 4.3.2.3 provides a detailed description of the conditions under which a channel can be shared between a pair of services.

5.5.2 Complexity of Shared Backup Path Protection Routing

The routing algorithm for SBPP services finds a primary path and a diversely routed backup path such that the total cost of channels used is minimized. Even with default SRGs, the problem of routing shared backup path-protected services to minimize the cost of new channels is NP-complete [107] as illustrated in Section 5.8.2. Chapter 6 examines heuristics to solve this routing problem.

5.6 Routing ILP

Many routing and network design problems in mesh networks can be formulated using the framework of Integer Linear Programs (ILPs) [224]. Such ILPs can be solved using off-the-shelf ILP solvers such as CPLEX [161]. Some expertise in the theory of linear programming is required to formulate the ILP and tune the ILP solvers to enable an efficient solution. ILP formulations for path routing and recovery problems in mesh networks have been presented in [93, 164, 258].

There are several advantages of using ILP solvers to solve routing problems. After a routing problem has been formulated as an ILP, ILP solvers often provide optimal solutions, and when not providing optimal solutions, they are able to provide a measure for how far from optimal the solutions are. New constraints on the routing problem can be incorporated as long as they can be expressed in the framework of ILPs as linear constraints or as modifications to the objective function. The drawbacks of ILPs include: (a) ILPs often do not scale with the size of the problem, (b) they often need nondeterministic time to provide feasible solutions, (and for this reason the usage of ILPs for online routing is uncommon), (c) the addition of constraints to the routing problem is often not easy, and (d) some expertise is required to frame the ILP, and tune the solver to provide efficient solutions for the specific routing problem. As a result, ILPs usually play a role in network planning and design, in offline routing situations, and in providing a benchmark to evaluate online routing algorithms.

5.6.1 ILP Description

This section provides the ILP formulations for routing problems in mesh networks discussed in earlier sections. Given a network and a set of services, the ILP routes primary paths, and backup paths (when required). We distinguish between two schemes to allocate channels to backup paths as discussed in Section 3.1. In the first scheme, the channels on the backup path are not assigned (capacity is reserved for backup paths but channels on the backup path are selected during recovery after the failure), and in the second, the channels on the backup path are preassigned. We outline the differences between the two schemes in the ILP formulation where applicable. For each demand, the ILP formulation selects the primary path from a predefined set of possible paths,[3] and backups are determined to be diverse from their primary path. We assume that all SRGs are of the default type (*Type a*). Although this formulation may lead to suboptimal solutions if the set of possible primary paths is not appropriately constructed, it nevertheless allows us to pin down the primary paths, and has the advantage of being less complex and faster to solve.

---

[3]The set of possible paths can be generated by a K-shortest path algorithm and can be as large as desired.

Table 5.5: Notation for Inputs and Variables in the ILP

The following notation describes the inputs to the ILP.

- N: set of nodes

- E: set of undirected edges

- D: set of demands

- R^d: set of K candidate paths for demand d

- R_i^d : set of edges on the ith primary path for demand d, $1 \leq i \leq K$

- C_e: cost of link e, e member of E

- F: set of all failure scenarios

The following notation describes the variables in the ILP:

- P_k^d: binary, equal to 1 if the kth route is chosen as the primary path for demand d, 0 otherwise.

- N_i^d: binary, equal to 1 if node i appears in the backup path of demand d, 0 otherwise.

- B_e^d: binary, equal to 1 if backup path of demand d uses link e, e member of E. It is 0 otherwise. With *pre-assigned* channels (assuming fixed link capacities):

$$B_e^d = \sum_c B_{e,c}^d \qquad (5.1)$$

where we define $B_{e,c}^d$ to be 1 if backup path of demand d uses channel c, 0 otherwise.

- W_e: capacity on link e, e member of E (upper-bounded if pre-assigned channels.)

- $Z_{e,f}^d$: binary, equal to 1 if backup path of demand d uses link e upon failure scenario f, e member of E, f belongs to the set of all links of all primary paths for demand d. It is 0 otherwise. With *pre-assigned* channels (assuming fixed link capacities):

$$Z_{e,f}^d = \sum_c Z_{e,f,c}^d \qquad (5.2)$$

where we define $Z_{e,f,c}^d$ to be 1 if backup path of demand d uses channel c upon failure scenario f, 0 otherwise.

Table 5.5 illustrates the notation for the inputs and variables in the ILP. The objective function and constraints on the variables are illustrated in Table 5.6. A solution to the ILP determines the values of the variables which in turn determine the primary and backup paths (and channels used on the backup path in the case where channels are preassigned) for all services.

Table 5.6: Objective and Constraints in the ILP

Objective: Minimize the total cost of all channels used on the primary and backup paths of all services: $\sum_{e \in E} C_e W_e$

Constraints:
One primary path is chosen for each demand:

$$\sum_i P_i^d = 1, \forall d \tag{5.3}$$

Primary and backup paths must be diverse:

$$B_e^d + P_k^d \leq 1, \forall d, \forall k, \forall e \in R_k^d \tag{5.4}$$

Flow conservation equations that determine the backup path:

$$\sum_{e:e=(i,j)} B_e^d - 2N_i^d = \left\{ \begin{array}{ll} -1 & \text{if } i = s_d, t_d \\ 0 & \text{otherwise} \end{array} \right\}, \forall d, \forall i \tag{5.5}$$

The constraint on the link capacity must be satisfied:

$$\sum_d \sum_{i:e \in R_i^d} P_i^d + \sum_d Z_{e,f}^d \leq W_e, \forall e, \forall f \neq e \tag{5.6}$$

Constraints on the sharing variables:

$$Z_{e,f}^d \leq B_e^d, \forall e, \forall f \neq e, \forall d \tag{5.7}$$

$$Z_{e,f}^d \geq P_k^d + B_e^d - 1, \forall k, \forall e \neq f, \forall f \in R_k^d, \forall d \tag{5.8}$$

With pre-assigned channels the constraint on the sharing variables is:

$$Z_{e,f,c}^d \geq P_k^d + B_{e,c}^d - 1, \forall c, \forall k, \forall e \neq f, \forall f \in R_k^d, \forall d \tag{5.9}$$

$$\sum_c Z_{e,f,c}^d \leq 1, \forall e \neq f, \forall f \in R_k^d, \forall d \tag{5.10}$$

5.6.2 Implementation Experience

The ILPs discussed in the previous section were applied to various realistic telecommunications networks to benchmark the online routing heuristics described in Chapter 6 and offline reoptimization algorithms described in Chapter 10. The CPLEX 7.1 [161] solver was used to solve the ILPs. The primary path was routed on the shortest-cost path that permitted a diverse backup path. The backup path was selected by the ILP solver. The networks included between 10 and 25 nodes with an average node degree of 3. The number of services routed was between 100 and 300. All services were routed as SBPP services.

Table 5.7 illustrates the comparison between the different routing techniques. We find that the online routing heuristic is within 7% of the optimal solution produced by the ILP solver, and the reoptimization algorithm is within 3% of the optimal.

On a pentium PC, the CPLEX ILP solver takes a few hours to solve the ILP with unassigned channels on networks with 20–25 nodes, with average degree 3 and with about 200–300 SBPP

Table 5.7: Benchmarking the performance of routing heuristics, and reoptimization algorithms using ILP solutions. The percentage from optimal is illustrated for different networks and demands

Algorithm	Network A	Network B	Network C	Network D
CPLEX lower bound	0	0	0	0
Online routing algorithm	5.1	2.5	6.4	1.6
Reoptimization algorithm	0	0.5	2.75	1.2

services. The ILP with assigned channels scales worse, with the solver being able to solve the problem in a few hours for networks of size 10–15 nodes with average degree 3 and with 100 SBPP services.

5.7 Conclusion

This chapter examines the complexity of various routing problems in mesh optical networks. The shortest-cost routing problem is solved using Dijkstra's algorithm. Dijkstra's algorithm is widely implemented, and forms the building block for various routing heuristics. With an additional routing constraint, the shortest-cost routing problem is NP-complete. A well-known pseudo-polynomial algorithm to solve this problem was presented.

For diverse routing, SRG structures were classified into three types. With default-type SRGs, the diverse routing problem is solved using Suurballe's algorithm. When SRGs conform to constraints imposed by conduits, then we show that the diverse routing problem can be solved after a simple graph transformation. We prove that when SRG structures are completely general then the diverse routing problem is NP-complete.

For routing SBPP services, given the sharing rules described in Section 4.3.2, we show that the problem is also NP-complete.

An ILP formulation was presented for solving routing problems. The ILP formulation was solved using the commercial off-the-shelf solver CPLEX for various real telecommunications networks. The ILP formulation was used to benchmark the performance of online routing heuristics and reoptimization algorithms.

5.8 Appendix

5.8.1 Complexity of Diverse-Path Routing with General SRGs

The SRG-diverse routing problem is the following: given a mesh network with general SRGs, and given a new service request from node A to node Z, is there a feasible SRG-diverse route from A to Z?

The SRG-diverse routing problem is NP-complete [107]:

1. SRG Diverse Routing $\in$ NP. Given a pair of diverse routes from node A to node Z we can check that their edges do not share any SRGs.

2. We can apply the 3-SAT problem [128] to the SRG-diverse routing problem as follows: the 3-SAT Boolean expression consists of a set of clauses $C_1, C_2, \ldots, C_N$ where each clause is a disjunction of three literals, e.g., $C_1 = x_1, \underline{x}_2, x_3$ ($\underline{x}$ indicates not x). The problem seeks to find an assignment of Boolean values to the variables that satisfies all the clauses. Given a 3-SAT Boolean expression we derive a network topology as indicated in Figure 5.7. To each variable x, we associate two edges (each with available unassigned channels), one labeled x

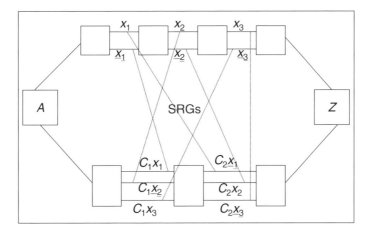

Figure 5.7: Network corresponding to the 3-SAT expression $C_1 = \{x_1, \underline{x_2}, x_3\}, C_2 = \{\underline{x_1}, x_2, \underline{x_3}\}$. (From [107], Figure 15. Reproduced by permission of © 2003 The International Society for Optical Engineering.)

and one labeled $\underline{x}$, as illustrated in Figure 5.7. To each clause $C = x, y, z$, we associate three edges (each with available channels), labeled C_x, C_y, C_z. There is an SRG defined between edge labeled C_x, and $\underline{x}$. There is an SRG diverse route between A and Z if and only if the 3-SAT expression is satisfied. If there is an SRG disjoint pair of routes between A and Z, then one of the routes would have to traverse the top part of the graph (through edges labeled x), and the diverse route would have to traverse the bottom part (through edges labeled C_x). If the route traverses through edge labeled x then the variable x is assigned the value 1, and if it traverses through the edge labeled $\underline{x}$, the variable x is assigned the value 0 (the route has to traverse one or the other edge). This assignment of Boolean values to the variables must satisfy each clause because, given clause $C = x, y, z$, the diverse route must traverse one of the edges C_x, C_y, or C_z. Say it traverses C_x, then the Boolean assignment to variable x must satisfy clause C (because of the way SRGs are defined). If there is a satisfying assignment of Boolean values to variables, then there is an SRG-diverse pair of routes. If variable x takes value 1, then the primary path traverses through edge x, and if x takes value 0, the primary path traverses through edge $\underline{x}$. For each clause $C = x, y, z$, the backup path traverses through one of the edges C_x, C_y, C_z, whichever is satisfied.

The difficulty of SRG-diverse routing arises because SRGs can be defined in arbitrary and complex ways which forces an algorithm to enumerate (a potentially exponential number of) paths in the worst case. As illustrated in Section 5.4.3, for some SRG configurations routing can be accomplished using existing diverse routing algorithms after performing appropriate graph transformations.

5.8.2 Complexity of SBPP Routing

The SBPP routing problem is the following: given a mesh network topology (with default SRGs), and a set of already provisioned SBPP services, and given a new service to be routed, is there a feasible primary and shared backup path from A to Z?

SBPP Routing Problem is NP-complete [107]:

1. SBPP Routing $\in$ NP. Given a feasible pair of primary and backup paths for an SBPP service, we can check that the primary and backup paths are edge-diverse, and that the rules of sharing are not violated on the backup path.

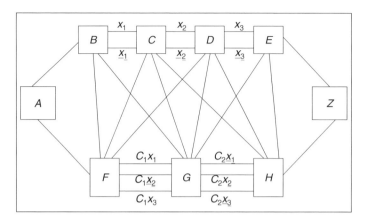

Figure 5.8: Network topology corresponding to the 3-SAT expression $C_1 = \{x_1, \underline{x}_2, x_3\}$, $C_2 = \{\underline{x}_1, x_2, \underline{x}_3\}$. (From [107], Figure 16. Reproduced by permission of © 2003 The International Society for Optical Engineering.)

2. We can apply the 3-SAT expression to the SBPP routing problem as follows: the 3-SAT Boolean expression consists of a set of clauses $C_1, C_2, \ldots, C_N$ where each clause is a disjunction of three literals, e.g., $C_1 = x_1, \underline{x}_2, x_3$ ($\underline{x}$ indicates not x). The problem seeks to find an assignment of Boolean values to the variables that satisfies all the clauses. Given a 3-SAT Boolean expression, we derive a network topology as indicated in Figure 5.8. This figure shows the network topology corresponding to the 3-SAT expression $C_1 = \{x_1, \underline{x}_2, x_3\}$, $C_2 = \{\underline{x}_1, x_2, \underline{x}_3\}$. Six services are already provisioned, $P_1 = \{Bx_1C, BFC_1x_1GC\}$, $P_2 = \{Cx_2D, CFC_1x_2GD\}$, $P_3 = \{Dx_3E, DFC_1x_3GE\}$, $P_4 = \{Bx_1C, BGC_2x_1HC\}$, $P_5 = \{Cx_2D, CGC_2x_2HD\}$, $P_6 = \{Dx_3E, DGC_2x_3HE\}$.

   To each variable x, we associate two edges (with available unassigned channels), one labeled x and one labeled $\underline{x}$, as illustrated in Figure 5.8. To each clause $C = \{x, y, z\}$, we associate three edges (with no available channels, and one shareable channel), labeled C_x, C_y, C_z. The set of provisioned SBPP services is defined as follows: For each clause $C = \{x, y, z\}$, there are three SBPP services defined, the first primary path has one edge $\underline{x}$, and the backup path has three edges, the middle one of which is edge C_x. There is a feasible SBPP service between A and Z if and only if the 3-SAT expression is satisfied. If there is a feasible SBPP service between A and Z, then the primary route would have to traverse the top part of the graph (through edges labeled x), and the backup route would have to traverse the bottom part (through edges labeled C_x). If the route traverses through edge labeled x then the variable x is assigned the value 1, and if it traverses through the edge labeled $\underline{x}$, the variable x is assigned the value 0 (the route has to traverse one or the other edge). This assignment of Boolean values to the variables must satisfy each clause because, given clause $C = x, y, z$ the diverse route must traverse one of the edges C_x, C_y, or C_z. Say it traverses C_x, then the Boolean assignment to variable x must satisfy clause C (because there is already a provisioned path whose primary path uses edge $\underline{x}$, and whose backup path uses edge C_x). If there is a satisfying assignment of Boolean values to variables then there is a feasible SBPP service from node A to node Z. If variable x takes value 1, then the primary path traverses through edge x, and if x takes value 0, the primary path traverses through edge $\underline{x}$. For each clause $C = x, y, z$, the backup path traverses through one of the edges C_x, C_y, C_z, whichever is satisfied.

In the common case, when diverse routes with available unassigned channels exist (and can be determined by standard diverse routing algorithms), a feasible SBPP service exists as well (because in the worst case, all unassigned channels can be used on the backup path with no sharing). However, the difficulty arises in the case where there are no diverse paths. This is because, in this worst-case, the routing of already provisioned SBPP services can lead to complicated sharing situations which makes it very difficult (short of exhaustive exponential enumeration of primary paths) to determine whether a feasible solution exists.

Chapter 6

Path Routing – Part 2: Heuristics

6.1   Introduction

Chapter 5 examined the computational complexity of various routing problems in mesh networks. It found that many routing problems are hard to solve optimally in polynomial time. In practice, however, a variety of heuristic approaches work well in solving these problems. In Chapter 6 we examine practical approaches to solve routing problems in mesh optical networks.

A routing heuristic tries to optimize certain metrics subject to the constraints of the network state, and service requirements. Depending on the type of service to be routed, the routing heuristic finds a working route for the service, and additionally, if the service is required to be protected, a diversely routed protection route is also determined for the service.

Finding optimal routes is often less important than finding a *good enough feasible* route. A routing heuristic needs to perform well in practical network scenarios while not necessarily providing optimal solutions. A robust routing framework allows the addition of new constraints, and accommodates the definition of new services without requiring a redesign for each change. Furthermore, a desirable property of the routing framework is that it can be tunable to produce better solutions if allowed to run for a longer period of time.

6.1.1   Operational Models: Centralized vs Distributed Routing

Figure 6.1 illustrates the interfaces of the routing subsystem with the other subsystems in a centralized routing implementation within a Network Management System (NMS). Figure 6.1 also illustrates the interfaces of the routing subsystem in a distributed routing implementation, that takes place in the Network Element (NE) as part of a distributed control plane used to provision services.

In an NMS, the topology manager constructs and tracks the network topology and states by polling the nodes, and upon receiving topology update events from the nodes. The NMS has complete state information for each node including the states of each port, all the services configured, and fine-grained bandwidth utilization for each link. Route computation is triggered for service provisioning, and usually service provisioning does not impose stringent requirements on the running time of the routing algorithm.

In a distributed control plane, the routing module interfaces to the topology manager that obtains network topology information from link-state routing protocols with traffic engineering extensions such as Open Shortest Path First (OSPF), Intermediate System–Intermediate System (IS-IS), or Private Network-to-Network Interface (PNNI). The route computation module also interfaces to the signaling module such as Resource Reservation Protocol (RSVP), Label Distribution Protocol (LDP)

Path Routing in Mesh Optical Networks   Eric Bouillet, Georgios Ellinas,
Jean-François Labourdette, Ramu Ramamurthy   © 2007 John Wiley & Sons, Ltd

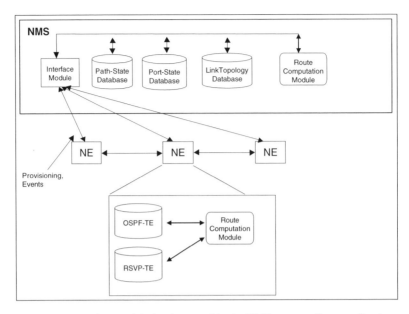

Figure 6.1: Route computation module implemented in the NMS as part of a centralized control plane and in the NE as part of a distributed control plane.

or PNNI-signaling. In a distributed control plane, route computation may be triggered for service provisioning, and it may also be triggered for service recovery upon a failure.

The routing heuristics described in this chapter apply to both centralized implementation in an NMS and distributed implementation in a network element.

6.1.2   Topology Modeling Example

Topology modeling for route computation in a mesh optical network, including the attributes of each link and each service, were discussed in Section 5.2. This section provides an example of such modeling for the network illustrated in Figure 6.2(a). The routing graph for the network in Figure 6.2(a) is illustrated in Figure 6.2(b). For each link with available bandwidth, a directed edge appears on the routing graph. The cost and delay on that directed edge is set to the cost and delay on the link. Figure 6.2(c) illustrates the topology database maintained by link-state routing protocols (such as OSPF) for the network in Figure 6.2(a). The topology database contains several attributes for each link including the cost of the link, the SRGs configured for the link, the total bandwidth of the link, the bandwidth currently available to route services, and the link-level protection available for the link. Such attributes of a link are used by the routing algorithm to find routes that satisfy the service requirements.

The routing graph has possibly multiple edges between a pair of nodes, as in Figure 6.2(b) between nodes A and B. The routing heuristics may convert such a graph to an ordinary graph (with at most one edge between each pair of nodes) by filtering the edges based on the service requirements.

6.2   Motivating Problems

We will motivate the routing heuristics with two routing problems.

- Constrained Shortest-Path Routing:

    Dijkstra's shortest-path routing algorithm [92] (described in Chapter 5) is widely implemented in link-state protocols like OSPF, and is the basic building block for Internet routing. The

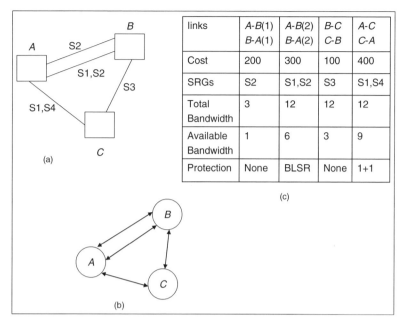

links	A-B(1) B-A(1)	A-B(2) B-A(2)	B-C C-B	A-C C-A
Cost	200	300	100	400
SRGs	S2	S1,S2	S3	S1,S4
Total Bandwidth	3	12	12	12
Available Bandwidth	1	6	3	9
Protection	None	BLSR	None	1+1

(c)

Figure 6.2: (a) Network example, (b) Graph model for the network, (c) Link-state topology database for the network.

objective for shortest-path routing is to minimize the cost of the links used on the path. In practice, apart from minimizing cost, additional non–trivial constraints are often imposed on a route. For example, the route must be of minimum cost while at the same time the total delay on the path must be bounded. Dijkstra's algorithm cannot solve the shortest-cost routing problem with such constraints. Finding the shortest-cost path, while at the same time satisfying certain constraints such as the delay constraint or the node-inclusion constraint, are NP-complete problems [128, 312]. Heuristic approaches are required to solve such routing problems.

The constraints imposed on the path for a service are usually of the following types:

- Min/Max Constraints: the minimum/maximum value of a certain metric over all the links of the path must be greater/less than a certain value. For example, the bandwidth constraint requires that each link on the path has a certain minimum bandwidth.

- Additive constraints: Constraints on metrics such as cost, delay and hop-count are additive over the links of the path.

- Inclusion/Exclusion Constraints: in such constraints, the path is constrained to include/exclude nodes and/or links of a certain type.

The sections that follow will examine heuristics for the above constraints.

- SRG-Diverse Routing:

Chapter 5 illustrated that the problem of finding diverse routes in a network is solved by various algorithms (including Suurballe's algorithm, and algorithms based on network flows [296]). Such algorithms are, however, not directly applicable to the SRG-diverse routing problem. Chapter 5 showed that the general problem of SRG-diverse routing is NP-complete. In this chapter we will examine practical heuristics to solve the SRG-diverse routing problem.

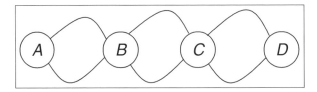

Figure 6.3: Exponential explosion of paths in a network.

6.2.1   Heuristic Techniques

The heuristics described in this chapter build upon Dijkstra's algorithm [92], the K-shortest paths algorithm [330] and Suurballe's algorithm [293]. Dijkstra's algorithm, and Suurballe's algorithm are described in Chapter 5, and the K-shortest path algorithm is described in Chapter 5, while its loopless version is described in this chapter.

In general, the heuristics utilize the following techniques:

- Path Enumeration: path enumeration explores a variety of paths in the network, seeking those that satisfy the constraints and service requirements. The number of paths between a pair of network nodes can be exponential in the size (measured in the number of links or nodes) of the network (as illustrated in Figure 6.3, where there are eight paths between A and D; these paths go through nodes A,B,C,D but differ in the links used), and therefore, not all the paths can be explored. However, a subset of paths can be chosen and explored. The K-shortest path algorithm enumerates paths in order of cost. There are other ways to explore paths, for example using search techniques such as breadth-first and depth-first search [86].

- Link Cost Function: in this technique, the routing algorithm modifies the cost of a link from the configured cost of the link to a function of attributes of the link and the attributes of the service to be routed. Then, a shortest-path computation is performed on the graph with the new cost function. An example of this technique is illustrated in the sharing-dependent routing heuristic for SBPP routing in Section 6.5.2, where, the cost of a link for backup path computation is redefined depending on whether that link is shareable, and then the routing algorithm uses the redefined cost.

6.3 K-Shortest Path Routing

The K-shortest path problem is a natural extension of the shortest-path routing problem in a network. The K-shortest path problem seeks to determine not just the shortest path, but K-shortest paths in order of increasing cost. Figure 6.4 illustrates K-shortest paths in a network. In this example, the four shortest paths between node A and node C are: ($A-B-C$ with cost 2), ($A-D-E-C$ with cost 4), ($A-D-B-C$ with cost 5), and ($A-B-D-E-C$ with cost 5). Various restrictions can be placed on the K-shortest paths. The most relevant one for mesh networks is that each of the K paths be loopless[1]. Further restrictions that are relevant in telecommunications networks, which have been studied extensively in the literature, include the restriction that the K paths be edge or node-disjoint [206]. Algorithms such as Suurballe's [293] or those based on network flows [14] for node or edge-diverse routing have been examined in Chapter 5. For the purpose of developing the routing heuristics in this chapter, we will use the K-shortest path algorithm with the only additional restriction that it produces loopless paths.

---

[1]A loopless path does not have repeated nodes. In telecommunications networks, paths with loops are always avoided.

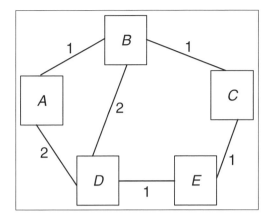

Figure 6.4: Example of a network used to find K-shortest paths between nodes A and C.

The K-shortest path problem has been widely studied since the 1950s. In the version of the problem where the paths are not required to be loopless the Epstein algorithm [116] achieves the best running time complexity of $O(M + N\log N + K)$ (where M is the number of links and N is the number of nodes in the network). When the paths are required to be loopless (which is the case in telecommunications networks), the best running time for directed graphs is $O(KN(M + N\log N))$ due to Yen's algorithm [330].

6.3.1 Yen's K-Shortest Path Algorithm

Yen's algorithm illustrated in Table 6.1 uses a shortest-path algorithm (such as Dijkstra's algorithm) as a subroutine. It begins by computing the shortest cost path between the source and the destination node. A deviation from a path is another path with the same source and destination, and having some number of initial nodes in common but deviating from the path at some node. Yen's algorithm works on the fact that the kth shortest path must be a deviation of the shortest paths 1 through $k - 1$.

In each iteration, the algorithm makes at most N invocations of Dijkstra's algorithm. Hence the algorithm performs KN invocations of Dijkstra's algorithm in the worst-case. Dijkstra's algorithm implemented using a Fibonacci heap has a worst-case running time of $O(M + N\log N)$. Hence Yen's algorithm scales as $O(KN(M + N\log N))$.

Several improvements to Yen's algorithm have been studied in the literature. Yen's algorithm has been improved by Lawler [195], but the worst case complexity has not been improved to date. [174] provides an algorithm for undirected graphs that has a running time of $O(N(M + N\log N))$.

In some networks, K-shortest paths do not provide topologically distinct paths as illustrated in Figure 6.5. In such cases, K-shortest diverse paths may be computed [206, 293] to provide a set of topologically diverse paths. Practical applications of the K-shortest path algorithm are examined in the following sections.

6.3.2 Constrained Shortest-Path Routing

Section 6.2 discussed various types of constraints on the shortest path. Routing with min/max constraints is achieved by topology filtering prior to applying a shortest-path algorithm. Topology filtering involves the removal of all links that do not satisfy the min/max constraint. For example, for a route with a certain bandwidth requirement, all links with lesser bandwidth are filtered from the topology prior to applying a shortest-path algorithm. Topology filtering can be applied for inclusion constraints that apply to all links/nodes on the path, and exclusion constraints that require the exclusion of certain links/nodes on the path.

Table 6.1: Yen's K-shortest path algorithm

Definitions:

- Consider a directed graph with N vertices $v_1, v_2, \ldots, v_N$.

- Let v_1 be the source and v_N be the destination for the k-shortest path computation.

- Let d_{ij} be the cost of the edge (v_i, v_j). Let M be the number of edges.

- Let A^k be the kth shortest path, $k = 1, 2, \ldots, K$ from v_1 to v_N.

- Let v_i^k be the ith node on the kth shortest path.

- Let A_i^k, be a deviation from A^{k-1} at the ith node on A^{k-1}. A_i^k follows the same path as A^{k-1} until the ith node of A^{k-1}. Then its $i^{th} + 1$ node is different from any path A^j $j = 1, 2, \ldots, k-1$ which also follows the same path as A^{k-1} until the ith node of A^{k-1}.

- The root R_i^k of A_i^k is the subpath of A_i^k that coincides with A^{k-1} until the ith node of A^{k-1}.

- The spur S_i^k of A_i^k is the subpath of A_i^k starting at the ith node of A_i^k and ending at the end node v_N.

Algorithm:

1. Initialize heap B.

2. Determine A^1 as the shortest path using Dijkstra's algorithm.

3. Iterate $K - 1$ times to determine A^i, $i = 2, 3, \ldots, K$.

   - Find all deviations of A^{i-1} and add each one to heap B.

   - Extract the minimum cost path from heap B as A^i.

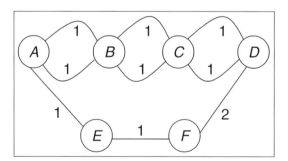

Figure 6.5: The first eight paths (each of cost 3) are not topologically diverse.

Table 6.2: Shortest path routing with constraints

---

1. Create the routing graph with topology filtering.

2. Run Dijkstra's algorithm and check if the shortest path meets the constraints. If so, return the path.

3. Find the next shortest-cost path P_i using the K-shortest path algorithm.

4. If P_i satisfies constraints, return the route.

5. If the number of paths found exceeds K, return NO ROUTE FOUND.

6. Iterate with step 3.

As presented in Chapter 5, shortest-path routing with an additional constraint on an additive path metric is NP-complete [312]. For example, if there is an additional delay constraint on the path, then the problem of minimizing cost while at the same time meeting a delay constraint is NP-complete. An exact pseudo-polynomial algorithm [168] for the problem was described in Chapter 5. In this section we examine an alternative approach based on path-enumeration as illustrated in Table 6.2. The heuristic in Table 6.2 enumerates paths in order of cost, and returns the first path that satisfies the constraints.

6.4 Diverse-Path Routing

In this section we illustrate heuristics for SRG-diverse routing. The simplest heuristic for SRG-diverse routing is illustrated in Table 6.3.

However, the above algorithm may fail to find a feasible path in topologies that contain a trap as shown in Figure 6.6. In this example, the shortest path $A-B-D-F$ does not have a link-diverse path,

Table 6.3: A simple heuristic for diverse routing

---

1. Find shortest path P.

2. Filter links with SRGs on P, and find shortest path B diverse from P.

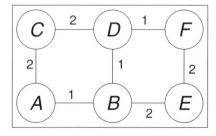

Figure 6.6: Trap topology for diverse path routing.

Table 6.4: A heuristic for diverse path routing

1. Iterate to find K shortest paths P_i, $i = 1, \ldots, K$.

   – Filter links with SRGs on P_i, and find shortest path B_i.

   If B_i exists return P_i, B_i.

Table 6.5: The distribution of K to find diverse routes in six telecommunications networks. For each value of K, the percentage of node-pairs that require that value for diverse-route computation is illustrated

K	Net1	Net2	Net3	Net4	Net5	Net6
1	95%	100%	94%	94.5%	100%	98%
2	5%		5%	5.5%		1%
3			0.25%			0.5%
4			0.25%			0.5%
13			0.25%			
16			0.25%			
K_{av}	1.05	1	1.13	1.05	1	1.03

Table 6.6: Distribution of K in a network with complex SRG configurations

Net7	1	2	3	5	7	9	13	17	21	29	33	41	49	65	81	113	129	145	161	K_{av}
	70.8	0.4	5.2	5.2	0.4	2.0	2.0	4.8	0.8	0.8	2.8	1.2	0.4	0.4	0.4	0.4	0.4	0.8	0.8	8.4

whereas a longer path $A-B-E-F$ has a diverse path $A-C-D-F$. One approach for the algorithm to avoid the trap is simply to try more than one primary path, until a diverse path is found. Such a heuristic is illustrated in Table 6.4.

The heuristic in Table 6.4 is parameterized by K which is the number of shortest paths that are explored. Table 6.5 illustrates the average value of K for finding diverse routes between different node-pairs for a variety of real telecommunications networks. We find that the average value of K is close to 1 for finding diverse node-pairs, and there are some node-pairs for which a larger K is required. Network 7 in Table 6.6 is an exception in which the average K to find diverse routes is more than 8 due to the presence of several complex SRG configurations.

6.4.1 Best-Effort Path Diversity

In some network topologies or under certain network conditions (for example when the network is congested with many links close to capacity), there may not exist diverse primary and backup paths in the network between several node-pairs. In such cases, it may still be desirable to provision protected services by allowing the primary and backup paths to violate the diversity requirement, i.e., the primary and the backup paths may have common SRGs. Upon a failure of the common SRG, recovery for the service fails since the backup path also uses the failed SRG. A measure of the degree of diversity between the primary and backup paths is the number of SRGs in common between them. Using this measure, the heuristic for best-effort diversity may seek to minimize the number of common SRGs between the primary and backup paths. This is achieved by the following steps: (a) after a primary path is found, the routing graph is modified by assigning large costs to the links with the SRGs on the primary path, and (b) a backup path is found on the modified graph.

6.5 Shared Backup Path Protection Routing

An SBPP service from node A to node Z is a pair of SRG-disjoint paths in the network where one of the routes is a working route, and the other is a backup route. Each link of the working route has dedicated capacity allocated to the service (and carries traffic under normal conditions). Each link of the backup route also has capacity dedicated to this service; however, the capacity dedicated on the backup path can be shared with backup paths for other SBPP services. For any two SBPP services s_1 (working route $= p_1$, backup route $= b_1$) and s_2 (working route $= p_2$, backup route $= b_2$), b_1 and b_2 can share an SRG if p_1 and p_2 are SRG-disjoint. The above sharing condition ensures that all SBPP services can be recovered after any single SRG failure. When any single SRG fails, all SBPP services whose working routes include the SRG, can be routed on their backup routes without any contention for capacity.

As shown in Chapter 5, even with default SRGs, the problem of routing SBPP services to minimize cost is NP-complete. The following sections present heuristics for this routing problem.

6.5.1 Sharing-Independent Routing Heuristic

The simplest heuristic for routing SBPP services is identical to SRG-diverse routing described in Section 6.4. After diverse working and backup routes are found, channels are shared on the backup path with other services where possible. This routing heuristic is oblivious to possibilities of sharing on the backup path; it requires only information that is needed for diverse routing.

However, we find a significant improvement in capacity utilization by utilizing the heuristic outlined in Section 6.5.2.

6.5.2 Sharing-Dependent Routing Heuristic

This heuristic exploits the intuitive definition of the cost of a backup path for an SBPP service. A meaningful notion of the cost of an SBPP service p whose working route is w and backup route is b is the sum of the costs of the links in w and costs of the unshared links in b. This definition sums up the costs of *new* capacity allocated to the service.

Given a primary path, the heuristic defines a cost function for backup path computation as follows:

- Filters links with SRGs on primary path.
- If a link is shareable, sets its cost to 0 (Chapter 7 examines how the cost can be set so as to achieve a trade-off between network cost and recovery time).
- If a link is not shareable, sets its cost to the link cost.
- A shortest path based on the new cost function is selected as a backup path.

The heuristic for routing is illustrated in Table 6.7.

Given a primary path, determining if a link can be shared on its backup path with some other service requires knowledge of all services provisioned in the network. Such information may not be available to the routing algorithm implemented at a node, whereas such information may be available at the NMS. Hence, while the sharing-independent routing heuristic for routing SBPP services can be implemented at a node, the sharing-dependent routing heuristic is suitable for implementation at the NMS.

Table 6.7: A heuristic for routing SBPP services

1. Apply the Sharing-Independent Routing Heuristic and save feasible diverse paths $(w(0), b)$.

2. Determine K-shortest paths $w(1), w(2), \ldots, w(K)$.

3. For each path $w(i)$ of the $K + 1$ choices for the working path:

   – For a link with SRG on working route, set cost to infinity.
   – For a shareable link, set cost to 0.
   – Otherwise, set cost to link cost.
   – Determine the shortest path, this is the backup route.
   – If the cost of the working and backup route is the best found so far save it as the best working and best backup routes.

4. Return the best working, backup paths.

Table 6.8: Capacity performance comparison between two heuristics for routing SBPP services

Network	Heuristic	Primary capacity	Backup capacity	Total capacity	Normalized total capacity
Net1	Heuristic1	574	406	980	100
	Heuristic2	546	520	1066	109
Net2	Heuristic1	2612	952	3564	100
	Heuristic2	2498	1576	4074	114
Net7	Heuristic1	90.1	92.9	183	100
	Heuristic2	78.9	132	210.9	115

Table 6.8 compares the two heuristics for routing SBPP services on three representative telecommunications networks with realistic service demands. It illustrates that from a capacity perspective the sharing-independent routing heuristic requires about 9–15% more total capacity than the sharing-dependent routing heuristic to route the same set of services.

6.6   Routing Preemptible Services

A preemptible service is a low-priority unprotected service whose channels may be used for restoring other (SBPP) services upon an outage. Hence, for preemptible services, channels that are on backup paths of SBPP services can be used. However, if such a channel is not available, then the preemptible service may use any available bandwidth on a link. The heuristic for routing preemptible services assigns a cost function to links as follows:

- If a link has an available shared channel, then the cost of that link is set to 0 (Chapter 7 examines a cost function that trades off capacity and path length).

- Otherwise, the cost of the link is the link cost.

A shortest path computation with the above cost function provides a good heuristic to route preemptible services.

6.7 General Constrained Routing Framework

In this section, we present a general constrained routing heuristic framework that ties up the ideas from the earlier sections.

When using the K-shortest path algorithm as the building block for the heuristics, it is difficult to know a priori the value of K to use for a given topology. Our experience on real network topologies indicates that most node-pairs require a small value of K to find diverse routes. When many different path constraints have to be satisfied, it is desirable to have the heuristic explore a multitude of paths, by setting a large value for K. On the other hand, in the interest of minimizing the algorithm running time, it is desirable for the algorithm to stop upon finding one (or a small number of) feasible routes that satisfy the constraints. With this objective, a simple enhancement is to iterate on primary paths until the first primary path is found with a feasible fully diverse backup path, and which satisfies any additional constraints. Thereafter, a small number of extra iterations are made over the primary paths. The pseudo-code for such a heuristic is illustrated in Table 6.9.

Table 6.9: A general routing heuristic

---

1. Run Suurballe's algorithm with default SRGs to find diverse paths. If SRG-diverse paths are found and the constraints are satisfied save as a feasible solution.

2. $d = 0$.

3. fully diverse solution found = false.

4. for ($i = 1$ to K_1)

   - P_i = next shortest primary path in the primary graph.
   - If (P_i does not satisfy constraints) then continue.
   - Determine backup routing graph based on P_i.
   - for ($j = 1$ to K_2)
     * B_j = next shortest path in backup routing graph.
     * If (B_j does not satisfy constraints) continue
     * if (B_j and P_i are fully diverse) then fully diverse solution found = true
     * if B_j and P_i is best solution so far, then save and break
   - if (fully-diverse-solution-found) $d = d + 1$
   - if ($d > d_1$) break;

5. Return best solution.

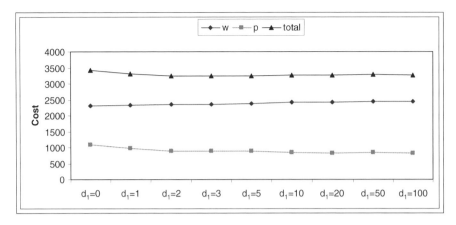

Figure 6.7: Capacity performance as d_1 is varied for a fixed $K_1 = 100$.

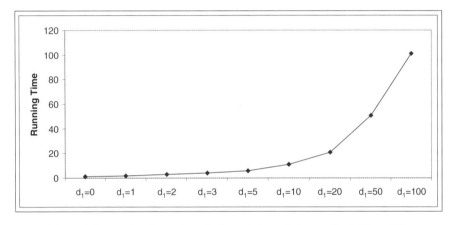

Figure 6.8: Running time of heuristic as d_1 is varied for a fixed $K_1 = 100$.

K_1 and d_1 control the iterations on the primary path, and K_2 controls the iterations on the backup path for a given primary path. Iterations on the primary path are needed for two reasons: (a) for satisfying the SRG-diversity constraint, and (b) for satisfying other constraints on the primary path. Iterations on the backup path are needed to satisfy constraints on the backup path.

If the kth ($\leq K_1$) shortest primary path has a fully diverse backup path, then at most $k + d_1$ primary paths are explored. If no fully diverse path exists until K_1 primary paths are explored, then the best solution, with diversity violations, is returned. The value of K_1 is selected to solve the SRG-diversity problem, so that fully diverse routes between all node-pairs can be found on all topologies. Our experience is that, on average, $1 + d_1$ primary paths are explored. In the worst case, K_1 primary paths are explored.

6.7.1   Implementation Experience

In this section we examine the performance of the above heuristic on a variety of networks for SBPP services, and determine the range of values for tuning the parameter d_1. As a benchmark for comparison, we use the case when the heuristic explores 100 primary paths (for example by setting $K_1 = 100$, and $d_1 = 100$). We compare this benchmark with the case when we vary d_1 from 0 to 100. Figure 6.7 illustrates the impact of the tuning parameter d_1 on the total capacity. We find that good feasible solutions are found for low values of d_1, when d_1 is less than 5. As d_1 increases, the heuristic selects solutions that have unusually long primary and/or backup paths, because they have the lowest combined primary and backup path cost after taking sharing into account. However, those solutions impact future demands and in the long run, they increase the total cost, which is why lower values of d_1 tend to perform better. Figure 6.8 illustrates the running time of the heuristic as d_1 is varied. As expected, the running time of the heuristic is linear in d_1.

The above data provides valuable insights into tuning the parameters of the heuristics. We find that good feasible solutions are found for the SBPP routing problem even when the number of primary paths explored is small, and hence the heuristic can be made to run faster without compromising its performance.

6.8 Conclusion

This chapter examined practical heuristics to solve routing problems in mesh optical networks. While many important routing problems such as constrained shortest-path routing and SRG-diverse routing are difficult to solve optimally, it demonstrated heuristic techniques that work well in practice. The heuristics are built with the algorithm for finding K-shortest paths as a basic building block. Results from the implementation of such heuristics on a variety of telecommunications networks are presented.

Chapter 7

Enhanced Routing Model for SBPP Services

7.1 Introduction

We have seen in earlier chapters that compared to dedicated backup path protection, shared backup path protection allows considerable saving in terms of capacity required [107]. In addition, the backup resources can be utilized for lower-priority preemptible traffic in normal network operating mode. However, recovery is slower than dedicated backup path protection, essentially because it involves signaling and path set-up procedures in order to establish the backup path when recovering from failures. In particular, we note that the recovery time will be proportional to the length of the backup path and the number of hops, and if recovery latency is an issue this length must be kept under acceptable limits. On the other hand this constraint may increase the cost of the solutions, as it is sometimes more cost-effective to use longer paths with available shareable capacity than shorter paths where shareable capacity must be reserved.

This is known as a bi-criteria optimization design (discussed in Section 4.6), in which the minimization of one objective occurs at the expense of a second. Experts in the field have experimented with various algorithms for solving this conundrum in the context of shared backup path protection. One possible method is to use a constrained shortest path routing technique that finds a minimum cost backup path that maximizes sharing opportunities, while remaining within a predefined bound such as a maximum number of hops, or a maximum length. The disadvantage of this approach is that it does not minimize the length of backup paths that are within the predefined bound. More importantly, it may fail to find a backup path if the bound is chosen too aggressively. For these reasons, researchers have explored methods that assign proper weights to the two objectives in order to achieve the desired compromise [55, 97, 241, 323, 325]. For instance, in [241, 325] the authors propose an Integer Programming (IP) formulation that takes these issues into consideration. They introduce a parameter $\epsilon \leq 1$ in the objective function in order to assign less weight to the cost of the backup path compared to the primary path. This forces the IP to prefer solutions that have longer backup paths than primary paths. They furthermore assign a weight $\mu \geq 0$ to shareable channels that expresses the penalty inflicted on the recovery time, even though shareable channels are virtually free. A similar approach using a shortest path routing algorithm instead of an IP was proposed in [55]. The reader should note that in [55], in the rest of this chapter, and throughout this book, we

Path Routing in Mesh Optical Networks   Eric Bouillet, Georgios Ellinas,
Jean-François Labourdette, Ramu Ramamurthy   © 2007 John Wiley & Sons, Ltd

Table 7.1: Integer programming formulation of shared backup path protection in dual link cost networks

- minimize:

$$\sum_{ij} w_{ij} + \alpha \sum_{kl} y_{kl} + \epsilon \sum_{mn} z_{mn}$$

- such that:

$$\sum_{ij} w_{ij} - \sum_{ki} w_{ki} = a_i$$
$$\sum_{ij} x_{ij} - \sum_{ki} x_{ki} = a_i$$
$$m_{ij,kl}(x_{kl} + w_{ij}) \leq y_{kl} + 1$$
$$x_{kl} \leq y_{kl} + z_{kl}$$

use parameters ϵ and α that correspond to parameters μ and ϵ respectively, as they were described above. In this chapter, we discuss these approaches according to the IP formulation described in Table 7.1.

The problem formulated in Table 7.1 is solved whenever we provision a new demand. In the formulation, w_{ij} is a binary variable equal to 1 if the working path uses link (i, j) and 0 otherwise; similarly x_{ij} is a binary variable equal to 1 if the backup path uses link (i, j) and 0 otherwise. Variable y_{ij} is a binary variable equal to 1 if a new backup channel is required on link (i, j) for the backup path, and 0 otherwise; similarly z_{ij} is a binary variable equal to 1 if link (i, j) is used for the backup and has a channel that can be shared with the channel of an existing demand, and 0 otherwise. The first two constraints represent the flow conservation constraint at node i for the working and the backup paths respectively. The right-hand side of the flow conservation constraints, a_i, is given and is equal to 1 if i is the source of the new demand, -1 if i is its destination, and 0 for all other nodes. Finally, $m_{ij,kl}$ is a given matrix value, equal to 2 if $(i, j) = (k, l)$, equal to 1 if the link (k, l) has no shareable channels available when the working path traverses link (i, j), and equal to 0 otherwise. One can verify using these values that according to the third constraint, the working and backup paths cannot traverse the same link, and that $y_{kl} = 1$ if $m_{ij,kl} = 1$, $w_{ij} = 1$ and $x_{kl} = 1$, and 0 otherwise. The objective is to minimize the total number of working links, plus the number of nonshareable backup channels weighted by $\alpha < 1$, plus the number of shareable backup channels weighted by ϵ. The values of α and ϵ can be adjusted to achieve the desired compromise between capacity requirements and backup path lengths.

We explore these trade-offs and experiment with an algorithm-centered metric by varying the weight put on the solution's cost versus the average backup lengths while selecting a primary-backup pair from a set of candidate routes. We assess the effect of the metric on these two contradicting objectives and identify the metric settings that achieve the desired compromise. We first introduce the routing metric, and then describe the algorithm used in our experiments to illustrate the effect of this metric.

7.2   Routing Metric

The quality of the routing algorithm and its ability to achieve the desired outcome will depend on how accurate is the routing metric used in expressing the quality of the solution. In this case it is both the minimization of the actual cost of provisioning a given demand set (combination of hardware, ownership and operational costs), and the minimization of the average path length of the solution. Because of the difference in units used to express the two terms, they must be assigned proper weights before they can be combined. As described in this section, the accuracy of the routing metric is a function of these two weights.

We define the length of a path as the sum of the predefined weights of the edges that constitute it. The metric or policy used for weighting the edges is different for primary paths and backup paths. For primary paths the weight of edge e is the real cost c_e of using the edge. For the backup path it is a function of the primary path. A backup edge e is assigned:

- infinite weight if it intersects with an SRG of the primary path.

- weight $w_e = \alpha c_e \leq c_e$ if new capacity is required to provision the path, and

- weight $s_e = \epsilon c_e \leq w_e$ if the path can share existing capacity reserved for preestablished backup paths.

The length of a primary and its protection path is then the sum of their respective path lengths. Quite evidently, the underlying idea here is to encourage sharing, whereby existing capacity can be reused for provisioning multiple backup paths.

The condition for sharing is that the backup paths must not be activated simultaneously, or in other words that their respective primaries must be pair-wise SRG-disjoint so that they do not fail simultaneously. The ratio $\epsilon = s_e/c_e$ can be adjusted for the desired level of sharing. For smaller values of ϵ, backup paths will be selected with the minimization of the number of nonshareable edges (weights w_e) in view, eventually leading to arbitrary long paths (as expressed in number of hops) that consist uniquely of shareable edges (weights s_e.) For larger values of ϵ, routing is performed regardless of sharing opportunities and backup paths will end up requiring substantially more capacity. This is demonstrated in Figure 7.1. The example depicts a demand (a, b) with its primary path routed on an eight-node network. It is assumed that all links have the same weight $c_e = 1$, and that the preexisting demand configuration is such that applying the routing metric described above on this

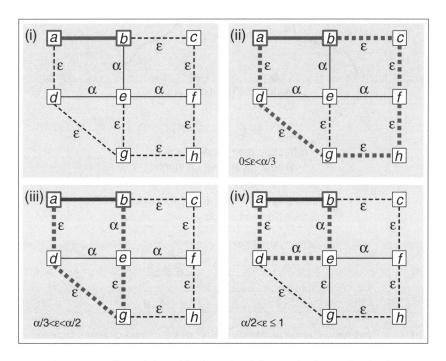

Figure 7.1: Effect of shareable channel weight ϵ on backup path selection.

configuration would result in the weights shown next to the links in Figure 7.1(i). Assuming for a moment that $\alpha = 1$, the example exhibits three ranges of ϵ leading to backup paths of different lengths, and different channel requirements.

Note that in the example above, the primary path is provided. However, a network design very often requires the computation of both the primary and the backup paths, and the routing metric should therefore differentiate between the cost of new primary channels and the cost of new backup channels. This is achieved by way of the parameter α. Typically, weight $w_e = \alpha c_e$ should be close to the real cost c_e of using the edge, that is $\alpha \lesssim 1$, since it indicates the absence of available shareable channels, and a new channel must be reserved. However, since it is likely that newly created backup channels will be shared in the future, it is sometimes advantageous to use a lower cost in order to encourage solutions that create new backup channels rather than primary channels.

This is illustrated in the examples of Figures 7.2 and 7.3. In the example of Figure 7.2, the primary path is given and the algorithm selects a backup path among two possible candidates paths p_1 and p_2. If when selecting the backup path, three new channels are required for using path p_1, and two new channels plus k shared channels are required for using path p_2, then p_1 is selected if $3\alpha < 2\alpha + k\epsilon$, that is if $k \geq \alpha/\epsilon$. The table of Figure 7.2 shows the minimum value of k for which p_1 is selected given different values of α and ϵ. Note that a smaller α is more likely to reduce the length of the backup path (given a fixed primary path), because it results in reserving new shared channels – useful for the future – which is similar to increasing ϵ. Figure 7.3 shows two nodes connected by a path p_1 which requires k new nonshareable channels (whether it is used as the primary or the backup path), and a path p_2 which consists of one shareable channel if p_1 is used as the primary path. In this

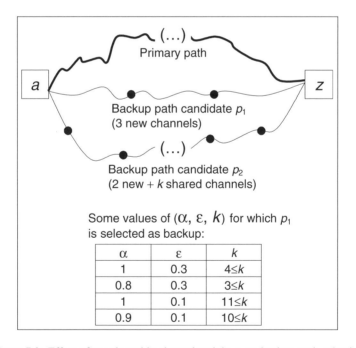

α	ϵ	k
1	0.3	$4 \leq k$
0.8	0.3	$3 \leq k$
1	0.1	$11 \leq k$
0.9	0.1	$10 \leq k$

Figure 7.2: Effect of nonshareable channel weight α on backup path selection.

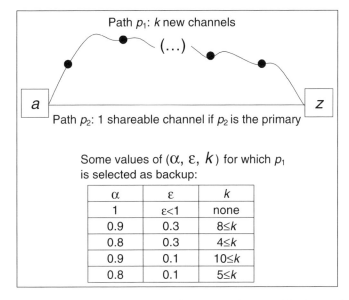

Figure 7.3: Effect of nonshareable channel weight α on primary and backup path selection.

example, path p_1 is selected as the backup if $1 + \alpha k < \epsilon + k$, that is if $(1 - \epsilon)/(1 - \alpha) < k$. The table of Figure 7.3 shows the minimum value of k for which p_1 is selected given different values of α and ϵ. Observe that in this case a smaller α is more likely to reduce the length of the primary path and increase the length of the backup path, because it results in creating shared channels which will be useful in the future.

7.3 Routing Algorithm

We now illustrate the impact of the routing metric in the context of a K-shortest path based algorithm. Recall that provisioning of lightpaths is performed in two steps: (1) computation of a primary-backup pair of routes, and (2) assignment of channels along the routes. Ideally the two steps are solved simultaneously and step (1) is optimized so that channel assignment in step (2) reuses existing capacity for backup paths. The K-shortest path algorithm solves the routing and channel assignment simultaneously, and optimally provided that K is chosen large enough. The algorithm is described in Table 7.2.

If the minimum cost is sought (maximum sharing), the value of ϵ in step 4(a)ii in Table 7.2, determining the cost of shareable protection channels, is set to 0. Otherwise, if shorter backup lengths and faster recovery times are desired, ϵ is set to a positive value. In the next section we first assume $\alpha = 1$ and study the effect of varying ϵ between 0 and 1. Extensive studies were performed for $\epsilon = 0$ in [107]. When the value of ϵ moves towards 1, we expect the lengths of primary and backup paths, as expressed in number of hops, to resemble that of dedicated backup path protection, though sharing is still implemented when available on the backup path and therefore the capacity required remains lower than that required for dedicated backup path protection. Finally, we experiment with α, and illustrate its effect on the routing algorithm.

Table 7.2: Shared backup path protection routing algorithm in dual link cost networks

---

1. For every edge e set weight to cost c_e of one channel in edge (cost of transponders, regenerators and optical amplifiers (OAs)).

2. Compute set P of k minimum-weight paths connecting node-pair $A-Z$, or all feasible paths if they are less than k of them.

3. Set $min\_weight = infinity$, and $\{p^*, q^*\} = INFEASIBLE$

4. For each primary path p in P:

   (a) Assign weight to every edge e:

      i. If e intersects SRG of primary path p, set weight to infinity.

      ii. If e has at least one channel that is shareable with p, set weight to $s_e = \epsilon c_e$.

      iii. Otherwise, set weight to $w_e = \alpha c_e$.

   (b) Using metric defined in step 4a, compute minimum-weight backup path q connecting node pair $A-Z$.

   (c) If q does not exist, continue at step 4, with next path p in P.

   (d) If min_weight < combined weight of paths p and q, then $\{p^*, q^*\} = \{p, q\}$ and min_weight = combined weight of paths p and q.

5. Return $\{p^*, q^*\}$

7.4 Experiments

7.4.1 Effect of ϵ

Figures 7.4–7.6 are representative samples of experiments with ϵ and α. Figures 7.4 and 7.5 summarize respectively the effect of ϵ on the average primary and backup path lengths and the ratio of protection (number of channels) to working capacity. The topology of the network used in these experiments is typical of an existing network. It consists of 50 nodes and 85 edges, with realistic demand traffic. We implement the algorithm of Table 7.2 using the K-shortest path algorithm presented earlier in Section 6.3. We choose the value K of the K-shortest path algorithm to be greater than the maximum number of paths between any pair of nodes. This is to guarantee that all possible paths are evaluated, and that the optimal solution to the IP given in Table 7.1 is found.

For this experiment the average backup length gradually decreases from 8 hops to less than 6 hops, a 25% reduction, as ϵ increases from 0 to 1. At the same time the protection capacity increases from 40 to 60% of the working capacity. The working capacity remains roughly constant across the experiments. Notice that in Figure 7.5 the ratio of the protection to working capacity is still less than 1 while it would be larger than 1 for dedicated backup path protection. Figure 7.6 illustrates the backup path histograms for two values of ϵ. The network used in this experiment is also representative of an existing network with 200 nodes and 300 edges. At $\epsilon = 0$ the figure exhibits a long-tail path length distribution (path lengths are expressed in number of hops) and even shows the existence of paths up to 60 hops. As ϵ increases, the width of the distribution decreases, and the maximum path

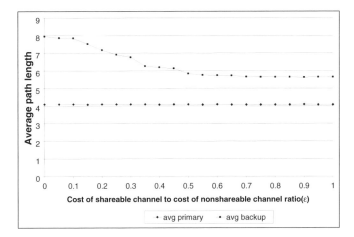

Figure 7.4: Effect of shareable channel cost to nonshareable channel cost ratio on average path length (50-node and 85-edge network). (From [55], Figure 3. Reproduced by permission of © 2002 The Optical Society of America.)

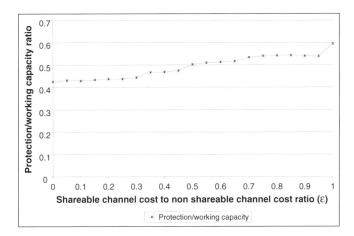

Figure 7.5: Effect of shareable channel cost to nonshareable channel cost ratio on protection to working capacity ratio (50-node and 85-edge network). (From [55], Figure 4. Reproduced by permission of © 2002 The Optical Society of America.)

length is reduced to 40 hops at $\epsilon = 0.4$. In the same time the protection capacity increases by 14%. The average path lengths and protection capacity increase show the same behavior as in Figures 7.4 and 7.5.

Figure 7.7 illustrates a typical range of values for ϵ that offers the best compromise between cost of solution and protection latency. Experiments indicate that values of ϵ in the range $0.2–0.4$ can be used to accommodate most of the optical networks encountered in practice. These findings are consistent with observations from [241].

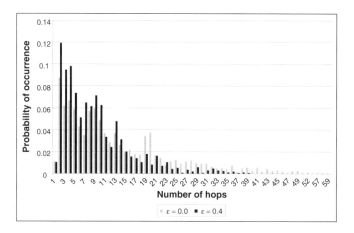

Figure 7.6: Effect of shareable channel cost to nonshareable channel cost ratio on backup length histograms (200-node, 300-edge network). (From [55], Figure 4. Reproduced by permission of © 2002 The Optical Society of America.)

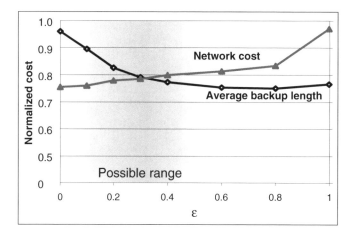

Figure 7.7: Range of ϵ offering good cost and protection latency trade-off.

7.4.2   Effect of α

Figure 7.8 illustrates the effect of α, the ratio of the weight representing non-shareable channels to the real cost of the channels. The example of the figure consists of a network of 11 nodes labeled from a to k, interconnected by 14 edges. One unit of demand is routed between nodes a and e, followed by one unit of demand between nodes b and j. Parts (i) and (ii) of the figure show the successive routing, using the algorithm presented earlier in this chapter, of demands (a, e) followed by (b, j), when $\alpha = 1$. If instead we chose $\alpha = 0.9$, we obtain the solution illustrated in parts (iii) and (iv) of the figure. We observe that in this example, using $\alpha = 0.9$ results in a solution that is 17% less expensive than the solution obtained with $\alpha = 1$. Note, however, that replacing demand (b, j) by demand (b, f) would reverse this observation. In fact we can show experimentally that if

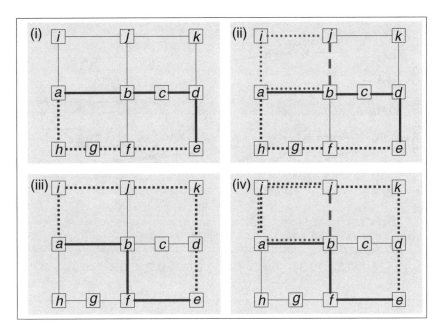

Figure 7.8: Effect of nonshareable channel weight α on routing algorithm.

we route any pair of demands between distinct nodes of degree 3 or more in this particular network, the average total number of channels required to route the demands (taking sharing into account) is about 9.6 for any values of α in the range 0.7−1.

Figure 7.9 illustrates the effects of α on the required capacity for various network topologies and respective demand sets taken from real case studies. As observed from the figure, the effect of α is highly dependent on the topology, and unlike the value of ϵ, its optimum value should be determined on a case-by-case basis. Note that these results are for the SBPP architecture in which protection channels are preassigned to the backup paths before failure. If the channels of the backup paths are selected during path recovery from pools of channels reserved for protection, then [241] observed that the effect of α was consistent over a wide variety of topologies.

7.5 Conclusion

In this chapter we observed that routing algorithms for the computation of shared backup path protection, with unrestricted sharing of protection capacity, achieve cost-effective solutions. However, they also result in longer backup paths and hence longer failure recovery times. This issue is addressed with an algorithm-centric metric that can be tuned to control the trade-off between sharing and backup path lengths when invoking the routing algorithm. The metric is based on a dual link cost-weighting function, which consists of assigning a different weight to a channel depending on whether it is reserved for the primary path or for the backup path. Noting that a channel already reserved for backup path protection costs virtually nothing if it can be shared with the new backup path, we assign to it a lower weight, equal to a fraction ϵ of the cost of the same channel, if it were to be reserved for the primary path. The value of ϵ can then be adjusted in order to keep the length of

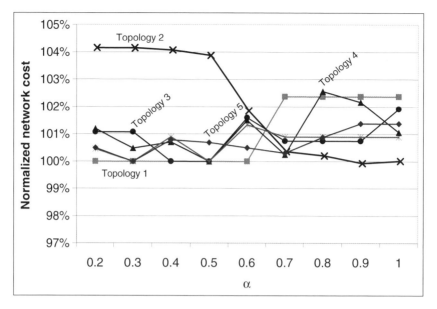

Figure 7.9: Effect of nonshareable channel weight α on various networks.

the backup path within the desired bounds. We experiment with the algorithm on real-life networks and show that assigning a small weight ϵ to shareable channels (instead of 0) results in a dramatic improvement in terms of backup path lengths. Furthermore, we note that the effect of a small ϵ is negligible on sharing for all the studied topologies. Evidence suggests that a value of ϵ in the range 0.2–0.4 achieves the best cost versus failure recovery time trade-off in most situations.

Chapter 8

Controlling Sharing for SBPP Services

8.1 Introduction

This chapter examines three topics relating to capacity sharing for SBPP services in mesh networks.

- *Routing with Express Links*: an express link (also known as a bypass link) can connect two nodes in the network that are not directly connected by a fiber link. Routing algorithms need to be enhanced to enable the efficient usage of such links on the primary and backup paths of SBPP services.

- *Limiting Sharing*: when routing SBPP services, it is possible for some shared channels to protect a disproportionately large number of services. Routing algorithms can be enhanced to limit sharing on shared channels.

- *Active Reprovisioning*: active reprovisioning is the provisioning of a new backup path for an SBPP service whose original backup path becomes unavailable due to a network failure or due to network maintenance activity. Reprovisioning the backup path enables SBPP services to survive a second failure which may occur during the time it takes to repair the first.

8.2 Express Links

Diversity of routes in a mesh network is defined using the notion of Shared Risk Groups (SRGs) as defined in Chapter 3. For example, all the channels that are multiplexed onto a WDM fiber link belong to the same SRG, since the failure of the fiber link simultaneously affects all the channels that are carried over that fiber. A set of channels between neighboring switches can, in general, belong to multiple SRGs. SRGs are configured by the network operator using information on the physical fiber plant of the network. Chapter 3 contains a detailed discussion of SRGs.

An express link consists of a set of channels that are routed over multiple concatenated DWDM fiber links as illustrated in Figure 8.1. At each intermediate node, the channels are not terminated at the switch but are connected directly from one DWDM system to the next. Express links are also called *glass-through* links since the channels are glassed through intermediate nodes. Chapter 8 of

Path Routing in Mesh Optical Networks   Eric Bouillet, Georgios Ellinas,
Jean-François Labourdette, Ramu Ramamurthy   © 2007 John Wiley & Sons, Ltd

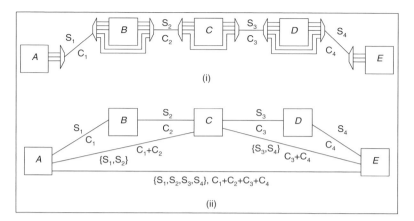

Figure 8.1: (i) Physical layout of a path with express links, (ii) Logical view of the network topology. (After [252], Figure 1. Reproduced by permission of © 2003 The Optical Society of America.)

[139] describes the motivation for express routes in transport networks, and examines capacity design and dual failure considerations in transport networks with bypass links.

Figure 8.1(i) illustrates the physical layout of an express link between node A and node C that is glassed through at node B, an express link between node C and node E that is glassed through at node D, and an express link between node A and node E that is glassed through at nodes B, C, and D. Figure 8.1(ii) illustrates the logical view of the network topology with assignment of SRGs and costs to the links. Each express link is assigned a set of SRGs that is the union of the SRGs belonging to each fiber link traversed by the express link. For example, in Figure 8.1 the SRGs assigned to the link $A-C$ are $\{S_1, S_2\}$, which is the union of the SRGs assigned to links $A-B$ and $B-C$. Such an assignment of SRGs ensures that two routes that are supposed to be diverse will not use the express link and any one of the underlying fiber links, since the failure of an underlying fiber link will result in the failure of both routes.

A glass-through link transports express traffic between a pair of nodes, and saves ports on the intermediate switches. However, the channels on each underlying fiber link (used in the express link) cannot be used to cross-connect to other channels at intermediate nodes. Express links are used when a portion of the traffic pattern is expected to remain unchanged over long periods of time. However, when traffic patterns change over time, then at each node, the flexibility of being able to cross-connect any channels on adjacent WDM fiber links using the optical switch is desirable.

8.2.1   Routing with Express Links

For routing SBPP services [252], we can divide the problem of routing into two parts: (a) route the primary path, and (b) route the backup path diverse from the primary path. As explained in Chapter 5, when the network topology can have SRGs that are arbitrarily defined, then the problem of diverse routing is NP-complete. Therefore, the selection of primary and backup routes is performed by a suitable heuristic that explores the space of primary and backup paths in an intelligent manner (as examined in Chapter 6). Channels on the primary path are dedicated to the service, and carry traffic. Channels on the backup path are shared with other services in such a manner as to ensure single-failure protection. The routing of each service will attempt to minimize the total cost of all channels used to route the service. A user-defined cost is assigned to fiber links that reflects the real cost of using a channel on that link. Assignment of the costs for the express links is discussed below.

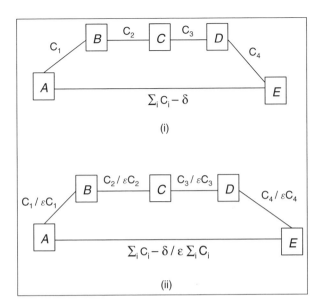

Figure 8.2: (i) Cost settings for primary path computation, (ii) Cost settings for backup path computation depending on whether the link is shareable or not shareable. (After [252], Figure 2. Reproduced by permission of © 2003 The Optical Society of America.)

Figure 8.2 illustrates the cost setting for an express link between node A and node E that is glassed through nodes B, C, and D. Figure 8.2(i) shows the cost of links for the primary path computation, while Figure 8.2(ii) shows the cost of links for the backup path computation depending on whether the link is shareable or not shareable. For routing of the primary path, the cost of each express link is set to be slightly less than the total cost of the underlying fiber links as illustrated in Figure 8.2(i). Such a cost setting ensures that the express link is always preferred over the path through the underlying links by a shortest-cost path routing algorithm. In general, if there are multiple tiers of express links as in Figure 8.1, then the cost of an express link between a node-pair is set to be slightly less than the cost of the shortest-cost path that uses the same underlying fiber links as the express link. For routing of the backup path, if the express link is not shareable, then its cost is set to be less than the sum of the underlying fiber link costs, and if the express link is shareable, then its cost is set to be a fraction of the sum of the costs of its underlying links.

8.2.2   Analysis and Results

We explore several different ways of setting costs to express links for routing the backup path of SBPP services. These approaches reflect the relative desirability of using a channel on the express link versus using channels on the underlying fiber links of the express link.

We simulated the routing behavior on a backbone network topology with express links, with several different demand patterns. The results of these experiments are reported in Table 8.1 for two demand sets. The measures of the service routing was the total number of ports used at all nodes, and the total bandwidth miles used on all the links. Table 8.1 illustrates the different cost functions for the express link, and the resulting backup path routing behavior. When the express link cost is less than the cost of the underlying fiber links (which is the case when it is shareable), the express link will always be preferred over the underlying fiber links. This is the case for the cost functions in rows 2, 3 and 4 of Table 8.1. Table 8.1 also illustrates that always routing backup paths on the

Table 8.1:   Different cost settings for routing backup paths on express links, and their resulting routing behavior. Results are reported for a representative topology with express links for two demand sets. (From [252], Table 1. Reproduced by permission of © 2003 The Optical Society of America.)

Express link cost	Backup path routing behavior	Switch ports		Bandwidth miles	
		Demand 1	Demand 2	Demand 1	Demand 2
Infinity	Always route on underlying links	824	3296	92.2K	368.8K
δ where $(\delta < \text{Min}(C_i))$	Always route on express links	846	3384	103.4K	413.8K
$\sum_i C_i - \delta$ where $(\delta < \text{Min}(C_i))$	Route on underlying links if at least one of them can be shared unless express link is shareable	632	2528	83.2K	332.8K
$\sum_i C_i - E$ where $E = $ average cost of underlying link	Route on underlying links if one (in some cases) or at least a certain number of them are shareable, unless express link is shareable	636	2544	83.4K	333.9K

express links does not perform well, and always routing on underlying links is better but it is not the best solution. The best performing approach routes backup paths mostly on the underlying links, but allows routing on express links if a shareable channel is available.

8.2.3   Express Links – Conclusion

We examined the routing of protected services in a mesh network with express links. To enable diverse routing, SRGs are assigned to the express link to be the union of all the SRGs on the underlying links. We examined the cost setting on the express links for routing the primary and backup paths. For routing primary paths, the express link is always preferred over the underlying links. For routing backup paths, the desirability of using express links depends strongly on the traffic demand. We find that biasing the routing of backup paths to mostly use underlying links while allowing express links to be used in some cases, performs well in terms of switch ports and bandwidth miles used.

8.3   Limiting Sharing

An SBPP service has a primary path and a diversely routed backup path. In one recovery architecture, backup routes are precomputed and protection channels on the backup path are preassigned at the time of path provisioning. Services whose primary paths are diverse can share channels on the backup path. Each shared channel protects a set of services which share that channel on their backup paths. If the routing algorithm does not discriminate between shared channels while routing a service, some shared channels may protect a large number of services (although the number of services protected by a shared channel averaged over all shared channels is small). For example, it is possible that the average number of services protected by a shared channel is 4, whereas there are several shared channels that protect more than 10 services. If shared channels protecting a large number of services fail, then those services are at risk upon a single failure on their primary paths. Upon a failure, the number of services that need to be recovered has a direct impact on the recovery time. By limiting the sharing on protection channels, it is expected that the recovery time performance will improve.

This chapter examines two approaches to limiting the number of services protected by a shared channel [253]. The goal is to eliminate the extreme cases of shared channels protecting a large number of services, while at the same time ensuring that the protection capacity does not increase significantly.

8.3.1 Example

Table 8.2 illustrates the distribution of the number of services protected by shared channels for a typical mesh network with 45 nodes and a demand of 80 services. The routing algorithm does not discriminate between shared channels, and as a result, there is a shared channel that protects 18 services, although, on average, a shared channel protects about 6 services.

8.3.2 Solution Alternatives

The following two approaches limit the amount of sharing on a channel on the backup path by influencing the channel selection procedure:

- Capping: with capping, a hard limit is imposed on the number of services using a shared channel. The routing algorithm considers only those shared channels which have not exceeded

Table 8.2: Distribution of the number of services protected by a shared channel. (After [253], Figure 1. Reproduced by permission of © 2003 The Optical Society of America.)

Services protected	Number of channels
1	10
2	12
3	6
4	6
5	18
6	7
7	1
8	8
9	4
10	4
11	2
12	0
13	1
14	4
15	2
16	2
17	0
18	1
Maximum services protected by a channel	18
Average services protected by a channel	5.98

the limit in the number of services that they protect. Work in [96] considers controlling the maximum sharing in mesh networks with SBPP services.

- Load-balancing: with load-balancing, the routing algorithm does not discriminate between shared channels during routing, however, during channel selection, the channel that protects the least number of links is selected. In this fashion, it is intended that each shared channel protects about the same number of links.

In both these approaches, the routing algorithm does not consider the amount of sharing on links; controlling channel sharing is considered only during channel selection. However, it is conceivable for the routing algorithm to take into account the amount of sharing on a link and penalize links with excessive shared capacity. This can be achieved by using the mechanisms outlined in Chapter 7, by influencing the cost of the links of the backup path based on the amount of shared capacity on a link. This approach is suited for the recovery architecture where shared protection channels are not preassigned to services, and controlling sharing can only be performed during route selection.

The following sections evaluate the two approaches outlined above.

8.3.3 Analysis of Capping

With capping, a limit is placed on the number of services that can use a shared channel. The routing algorithm considers only those channels that have not exceeded the limit in the number of services they protect. It is shown that a well-chosen sharing limit can be robust to the network topology and demand pattern.

We define,

- R: The ratio of the number of protection channels to the number of working channels in the network.

- L_{av}: Average number of services using a shared channel.

- L_{max}: Maximum number of services using a shared channel.

- h_w: Average number of hops on the working path (averaged over all services).

- h_p: Average number of hops on the backup path (averaged over all services).

By definition, $R = (h_p/h_w)(1/L_{av})$ i.e., R is inversely proportional to the average number of services using a shared channel. Also, $R \geq (h_p/h_w)(1/L_{max})$. Note that for $L_{max} = 1$, i.e., each protection channel can protect at most one service, we fall back to the DBPP case. Since the lower bound on R is inversely proportional to L_{max}, with a sufficiently large choice of L_{max}, changes in the value of L_{max} will cause small changes in R. In this sense, the specific choice of the sharing limit will not impact on the ratio of protection to working channels in the network as long as the sharing limit is sufficiently large.

In the recovery architecture without preassigned channels, the routing algorithm can use the average number of services using the shared channel on a link to weight the preference of that link on the backup path using approaches outlined in Chapter 7. For example, if P is the protection capacity on a link, and B is the number of backup paths using that link, then P/B is a measure of the average number of services using a shared channel on that link. P/B can be used to set the cost of that link in a routing algorithm to select a backup path.

8.3.3.1 Experiments

A set of experiments were performed to study the impact and sensitivity of the sharing limit on a variety of network parameters. The set of networks and traffic demands in the experiments are a mix of representative real networks and traffic, and randomly generated networks and traffic. Four representative real networks were considered: netA (45 nodes), netB (17 nodes), netC (50 nodes), and netD (100 nodes). Three randomly generated networks were also considered: net50 (50 nodes), net100 (100 nodes), and net200 (200 nodes).

8.3.3.2 Capacity

Figure 8.3 illustrates the ratio R of protection to working channels for different networks and for different values of the sharing limit. It is observed that as the sharing limit L_{max} increases, R decreases sharply at first, then decreases gradually, and finally remains flat. Most of the sharing gains are obtained when the sharing limit is below five services. In all networks, there are no incremental gains in protection capacity beyond the sharing limit of 10 services. The studies in [96] find that beyond a sharing limit of three, the capacity gains are marginal.

8.3.3.3 Backup Path Length

Figure 8.4 illustrates the average number of hops on the backup path as the sharing limit varies. The number of backup path hops directly influences the recovery time upon a failure, with a larger number of backup hops generally consuming more recovery time. It is observed that as the sharing limit increases, the average number of backup hops increases marginally. This is because, as the sharing limit increases, there are more opportunities for sharing, and the backup path may traverse a longer distance trying to use links with shareable channels. The routing algorithm can assign a cost model for shareable channels to achieve a trade-off between sharing and the length of the backup path.

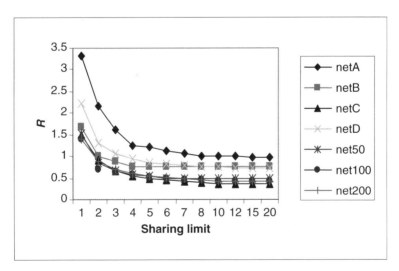

Figure 8.3: Protection to working capacity ratio vs sharing limit. (From [253], Figure 2. Reproduced by permission of © 2003 The Optical Society of America.)

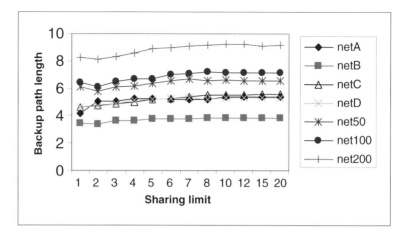

Figure 8.4: Backup path length vs the sharing limit. (From [253], Figure 3. Reproduced by permission of © 2003 The Optical Society of America.)

8.3.3.4 Double Failures

Figure 8.5 plots the impact of the sharing limit on the percentage of services that are recovered upon two simultaneous link failures for a given network. The values are averaged over all possible double link failures. Also plotted is the capacity requirement for each value of the sharing limit. It is observed that as the sharing limit increases, the required capacity decreases (due to a decrease in protection capacity because of better sharing); however, due to better sharing, there are more contentions for capacity upon double failures, and as a result, there is an increase in the percentage of services that are not recovered. The figure indicates that to achieve better resiliency against double failures, the sharing

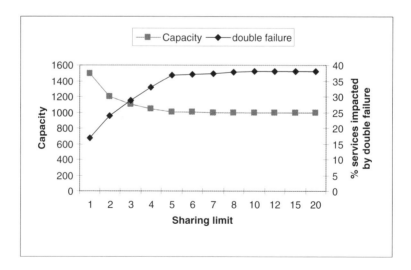

Figure 8.5: The impact of the sharing limit on double failures. (From [253], Figure 4. Reproduced by permission of © 2003 The Optical Society of America.)

limit needs to be decreased, and, consequently the capacity requirement increases. The experiments in [96] show that most of the gains in double failure recovery are achieved with a sharing limit of two or three.

8.3.4 Analysis of Load-Balancing

In load-balancing, the routing algorithm does not discriminate between shared channels during routing. However, during channel selection on the backup path, and given a set of channels that can be used on the backup path, various policies can be used for selecting the specific channel. Load-balancing considers the number of SRGs protected by a channel as the metric to select channels, and selects channels with the least number of SRGs protected. Another approach would be to select a channel at random among the set of allowed channels.

8.3.4.1 Capacity Requirement

Table 8.3 illustrates the ratio between the capacity required with load-balancing and the capacity required without load-balancing. This ratio is illustrated with and without imposing a sharing limit. It is observed that there is negligible and inconsistent differences between the capacity requirements with and without load-balancing, i.e., load-balancing on shared channels does not affect the capacity requirement.

8.3.4.2 Sharing Distribution

Table 8.4 illustrates the maximum number of services that are shared by some channels with and without load-balancing. It is observed that load-balancing does have a limiting effect on sharing, i.e., load-balancing does reduce the maximum number of services that are shared by some channels. However, the reduction in the maximum number of services shared by some channels is not deterministic, i.e., it is not possible to guarantee that a certain limit will not be exceeded. When there is a sharing limit, it is observed that load-balancing does not have any effect on the maximum number of services that are shared by some channels.

8.3.5 Limiting Sharing–Conclusion

Limiting sharing on channels eliminates the cases of shared channels protecting a large number of services. Setting a sharing limit appears to be robust across different network topologies and demands. Small values for the sharing limit achieve better recovery efficiency upon double failures, at the cost

Table 8.3: Load-balancing on protection channels and capacity. (After [253], Table 1. Reproduced by permission of © 2003 The Optical Society of America.)

Network	Without sharing limit	With sharing limit
netA	1.002	1.002
netB	1.000	0.992
netC	1.004	0.999
netD	1.004	1.003
net50	1.003	1.000
net100	0.996	0.995
net200	0.998	0.989

Table 8.4: Distribution of services on protection channels. (After [253], Table 2. Reproduced by permission of © 2003 The Optical Society of America.)

Network	Without sharing limit		With sharing limit	
	Without load-balancing	With load-balancing	Without load-balancing	With load-balancing
netA	9	6	8	8
netB	13	11	8	8
netC	17	12	8	8
netD	13	10	8	8
net50	13	9	10	9
net100	14	12	10	10
net200	17	17	10	10

of additional protection capacity. The sharing limit does not appear to have an impact on the length of the backup path.

Load-balancing limits the maximum number of services shared by some channels, but this limit is not deterministic. Load-balancing has a negligible effect on the capacity requirement, and load-balancing on shared channels does not provide quantifiable benefits in limiting sharing, especially if a sharing limit is already imposed.

8.4   Analysis of Active Reprovisioning

Active reprovisioning is the provisioning of a new backup path for a service whose original backup path becomes unavailable. Backup paths can become unavailable due to the following failure scenarios [249]:

- Primary channel failure resulting in recovery: An SBPP service switches to its backup, rendering the backup channels unavailable for other services that are sharing them. Figure 8.6 illustrates service P_1 that has a primary path $A-B-C$ and a backup path $A-D-E-C$ and service P_2 that has primary path $F-G$ and backup path $F-A-D-E-C-G$. In Figure 8.6(i), primary path P_1 switches to its backup, rendering the backup unavailable for primary path P_2.

- Backup channel failure: a backup channel fails due to an equipment (e.g. transceiver, amplifier, or fiber) failure. In Figure 8.6(ii), a channel on the backup path fails, rendering the backup unavailable for primary paths P_1 and P_2.

- Fiber failure: a fiber failure results in several services whose primaries use the failed fiber to switch to their backups simultaneously.

In addition to the above failure scenarios, work in [98, 143] considers span maintenance actions in the network that render the capacity of that span unusable for the maintenance duration. Such maintenance actions are similar in nature to a link failure if the maintenance action was preceded by switching the working services on that span onto their backup paths.

By reprovisioning a new backup path, the service can recover quickly from a double failure. However, since the network capacity is often provisioned only for single failures, successful reprovisioning is not guaranteed. Furthermore, the likelihood of a second failure during the time it takes to repair the first is often small. The sections that follow analyze the potential benefits of active reprovisioning.

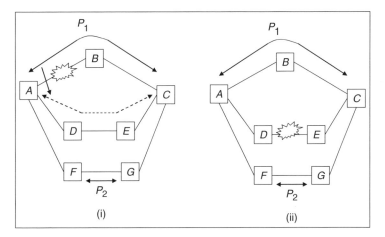

Figure 8.6: (i) Failure on a primary path renders the backup path unavailable for a second primary path, (ii) Failure on the backup path renders both primary services unprotected.

8.4.1 Evaluation of Active Reprovisioning

The performance of active reprovisioning was evaluated on three representative telecommunications networks: Network 1 (45 nodes, 75 links, 72 services), Network 2 (17 nodes, 26 links, 102 services) and Network 3 (50 nodes, 88 links, 300 services).

8.4.1.1 Number of Active Reprovisioning Requests

Table 8.5 illustrates the average number of services whose backups become unavailable after three different failure events. For backup channel failure, the number of services that need to be actively reprovisioned is equal to the number of services that share a protection channel. For a primary channel failure resulting in the service being switched to its backup path, the number of services that need to be actively reprovisioned is equal to the number of backup hops multiplied by the number of services that share a protection channel. For fiber failure, the number of services that need to be actively reprovisioned is equal to the average number of services on a fiber, multiplied by the average number of backup hops, multiplied by the average number of services that share a protection channel.

8.4.1.2 Success Rate of Active Reprovisioning

Table 8.6 illustrates the average percentage of services that are successfully reprovisioned after each failure event (assuming that network capacity is provisioned only for single failures). Also included in the table is the percentage of services for which primary and backup paths are not fully diverse after reprovisioning.

Table 8.5: Average number of services eligible for active reprovisioning after three different failure events. (From [249], Table 1. Reproduced by permission of © 2003 The Optical Society of America.)

Event	Network 1	Network 2	Network 3
Backup channel failure	4	2.2	4.2
Primary channel failure	13.6	7.6	25.1
Fiber failure	17	55	192

Table 8.6: Average number of services successfully reprovisioned after three different failure events. (From [249], Table 2. Reproduced by permission of © 2003 The Optical Society of America.)

Event	Network 1		Network 2		Network 3	
	% success	% diversity violation	% success	% diversity violation	% success	% diversity violation
Backup channel failure	96	28	90	23	93	11
Primary channel failure	76	19	85	38	90	11
Fiber failure	71	20	58	28	58	15

Table 8.6 shows that after a backup channel failure, 90% of the services can be successfully reprovisioned; after a primary channel failure, about 75% of the services affected can be successfully reprovisioned; and after a fiber failure about 60% of the services can be successfully reprovisioned. The success rate is directly impacted by the number of services that need to be reprovisioned. Work in [98] considers the problem of designing the network, so that it is guaranteed that the network is immune to maintenance-related actions.

8.4.1.3 Likelihood of Failure without Active Reprovisioning

Active reprovisioning is beneficial only when a second failure occurs while the first is being repaired. The probability of this event is computed as shown in the following illustrative example.

Assume that the first failure is the failure of a channel on the primary path resulting in the recovery of a service which then results in 20 services becoming unprotected.

- Probability that a transceiver (that terminates a channel) fails in 1 hour[1] $= 1.5 \times 10^{-6}$.

- Mean Time To Repair (MTTR) for a transceiver failure = 4 hours.

- Number of services unprotected during the time to repair = 20.

- Average number of hops on the primary path of each unprotected service = 3.

- The total number of transceivers on the primary paths of unprotected services $= 3 \times 2 \times 20 = 120$.

From the above assumptions, the probability of a second transceiver failure impacting on an unprotected service is computed to be $120 \times 1.5 \times 10^{-6}$ which is approximately 0.001.[2]

8.4.2 Active Reprovisioning–Conclusion

Active reprovisioning is the provisioning of a new backup path for a service whose original backup path becomes unavailable. Backup paths become unavailable either due to unexpected failures or due to scheduled maintenance actions in the network [98, 143]. Assuming that the network capacity is designed for single-failure recovery, reprovisioning is practical for channel failures on primary/backup paths and not for fiber failures, due to the low success rate of reprovisioning upon fiber failure. Based

---

[1]Assuming that the Failure In Time (FIT) rate for the transceiver is 1500.

[2]In this calculation, we have ignored the risk of fiber cut, since the FIT rate of terrestrial fiber cut/km is approximated to be 50 and trans-oceanic fiber cut/km is 5. Assuming 100 km links, the FIT rate for fiber cut is of the same order of magnitude as that of a transceiver.

on MTTR (Mean Time To Repair) and failure rates, the probability of a second failure impacting on the same set of services that become unprotected due to a first failure is low. Hence, reprovisioning is practical when applied to services that have the highest service guarantees and for channel failures on the primary/backup paths. Chapter 12 discusses dual failures in more detail, in the context of service availability.

8.5 Conclusion

This chapter examines various aspects of mesh networks that impact on the deployment of SBPP services.

In a mesh network with express links, this chapter proposes extensions to the routing algorithm to enable efficient capacity usage.

While SBPP services enable efficient sharing of protection capacity, it is possible for some channels to protect a disproportionately large number of services. Limiting sharing on channels eliminates the case of having a channel protecting a large number of services. Load-balancing on shared channels does not provide quantifiable benefits in limiting sharing especially if a sharing limit is already imposed.

Active reprovisioning is the provisioning of a new backup path for a service whose original backup path becomes unavailable. Reprovisioning is practical when applied to services that have the highest service guarantees and for channel failures on the primary/backup paths.

Chapter 9

Path Computation with Partial Information

9.1 Introduction

In earlier chapters we have discussed various approaches used by optical network operators in order to guarantee service persistence in case of network failure. We have seen in particular that it is common for a carrier to reserve spare bandwidth on alternate paths called backup paths, so that a service affected by a failure along its primary path can be rapidly recovered using the reserved bandwidth on the backup path. Among the possible schemes for provisioning backup paths, dedicated backup path protection (DBPP) and shared backup path protection (SBPP) appear to be the preferred approaches in the context of mesh optical networks [55, 93, 99, 100, 164].

Until now we have assumed that shareability of protection channels was determined using deterministic approaches that require a detailed level of information proportional to the number of active lightpaths. Although this is not an issue for small size networks in the foreseeable future, these approaches are not practical for distributed route computation involving larger networks. On the other hand, distributed approaches that do not make use of shareability information require a significant amount of additional capacity compared to a centralized approach with access to complete shareability information [52, 240, 251]. In this chapter we experiment with several shared backup path route computation algorithms that use varying degrees of sharing information. We show in particular that even with less information, independent of the amount of traffic demand, it is possible using statistical methods to predict the shareability of backup channels with remarkable accuracy. In addition, we discuss a local distributed channel assignment scheme that is used in conjunction with the distributed route computation method to assign backup channels when provisioning the shared backup path protection. This channel assignment scheme can also be used to further optimize capacity usage in individual links, either upon occurrence of certain events, or at predetermined regular intervals. We observe that the probabilistic approach yields faster computation times with surprisingly no significant penalty in terms of capacity usage than a centralized approach using complete information.

Before we address the computational complexity of the shared backup path protection architecture, and attempt to derive comparisons with dedicated backup path protection, a review of the

Path Routing in Mesh Optical Networks   Eric Bouillet, Georgios Ellinas,
Jean-François Labourdette, Ramu Ramamurthy   © 2007 John Wiley & Sons, Ltd

two architectures is in order. Recall that in dedicated backup path protection the lightpath provisioning algorithm computes and establishes simultaneously the primaries and their backup paths. During normal operation mode, both paths carry the optical signal and the egress node selects the best copy of the two. It suffices that a primary and its backup path be SRG-disjoint to ensure that at least one path survives any single failure affecting all the channels in an SRG (see Chapter 3).

As in dedicated backup path protection, shared backup paths are predefined, except that the cross-connections along the paths are not created until a failure occurs. During normal operation mode the spare channels reserved for failure recovery are not used. Since the capacity is only reserved and not used until a failure occurs, the same backup channel can be shared to protect multiple lightpaths. There is a condition though that two backup paths may share a reserved backup channel only if their respective primaries are mutually SRG-disjoint, so that a failure does not interrupt both primary paths. Two paths, or their protection, are said to be *mutually compatible*, if they are not affected by the same failure. If not, they are *conflicting*.

There are two different policies to assign channels to backup paths [106, 187]. A failure independent strategy reserves and assigns the backup channels at the time of provisioning before failures occur. A failure dependent strategy (channel pooling) assigns the backup channels along a precomputed backup path after failure occurrence and relies on the recovery signaling mechanism to select the channels on each link along the path from a pool of reserved channels [89]. The latter relies on a proper channel-provisioning scheme to reserve enough channels on each link so that all lightpaths can be recovered for every type of single SRG failure. Although the difference between the two strategies may appear subtle at first, a comparison between the failure dependent versus the failure independent shared backup path protection will convince us of its importance as illustrated in Figure 9.1. The example consists of three demands (*AB*, *AC*, and *BD*), routed across a six-node optical network in

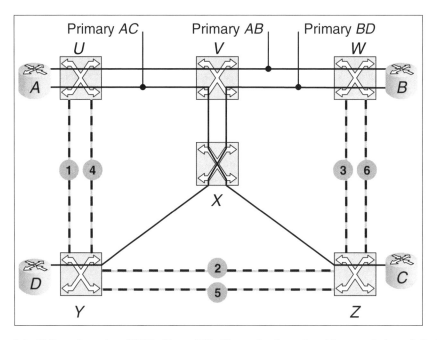

Figure 9.1: Failure dependent SBPP. (From [52], Figure 5. Reproduced by permission of © 2004 The Institute of Electrical and Electronics Engineers.)

such a way that every combination of primary lightpath pairs, but not all three primary lightpaths at once, can fail simultaneously. If a failure dependent shared backup path protection strategy is used, only two channels need to be reserved on the backup path, since at most two lightpaths will fail simultaneously. If, however, a failure independent strategy is used, we must reserve three backup channels on link (Y, Z) in order to accommodate all failure scenarios affecting links (U, V), (V, W) or (V, X), even though at most two of the three channels will be used at any time. The reason for this is that link (Y, Z) is traversed by the protection paths of all three demands. Since any combination of primary lightpath pair can fail, each of them must be preassigned a different backup channel in order to avoid contention.

Although more cost-efficient, the failure dependent shared backup path protection approach requires additional inter-node communication to agree on the channel assignment during failure recovery [102]. Instead, we prefer the failure independent approach, which has achieved sub-200 ms recovery times in large networks [18, 66]. However, the gain in failure recovery time requires that backup-to-channel lookup tables be determined at each node during shared backup path provisioning (when speed is less of an issue).

Now, consider the online problem of provisioning shared backup paths using a centralized route computation, assuming a failure independent strategy. Since this problem is proven to be NP-complete (see appendix in Chapter 5), if minimization of the total capacity usage (working plus protection) is sought [106], a possible approach is to enumerate a list of K minimum cost primary paths and for every one of them compute the corresponding minimum cost backup path and reserve the channels along that path. The route computation then returns the primary and backup path pair with the lowest combined cost. The cost of a pair is the cost of the channels along both paths, excluding the cost of (preexisting) shareable reserved channels along the backup path. Given a primary path, we compute the minimum cost backup path by:

1. Setting the cost of the links (SRGs) traversed by the primary path to ∞,

2. Setting the cost of links with shareable channels to a constant $\epsilon \ll 1$,

3. Running a shortest path algorithm using the modified link cost metric.

Steps (1) and (2) ensure that the primary and backup paths respectively are SRG-diverse, and that the minimum cost backup path is found using shareable reserved channels whenever possible. In the sections that follow we are interested in step (2), which consists of identifying shareable reserved channels. We show in particular that the time complexity of this operation, if deterministic, is proportional to the total number of primary and backup channels reserved for existing lightpaths, and thus does not scale well when the number of lightpaths established in the network becomes large. We then present a probabilistic approach to execute this operation with a certain probability of accuracy. We show that by trading a deterministic statement for a probabilistic statement, the operation can be made independent of the number of reserved channels. The benefits of this substitution are twofold. First, it allows the reduction of the backup path computation time, and second it allows the reduction of the amount of information necessary to compute the paths, with no penalty or small penalty in terms of capacity efficiency. The probabilistic approach computes the backup paths, but, unlike the deterministic approach, it does not provide the channels along the path, and this assignment must be done separately as provisioning of the path takes place on a link-by-link basis. We show that this backup channel assignment operation is equivalent to a graph-coloring problem. In particular, we show how a first-fit based assignment can be easily improved using a graph-coloring algorithm. The result is a routing architecture that is more comparable to the failure dependent channel allocation policy in terms of computation and information complexity, but still maintains the failure recovery latency of the failure independent strategy by preallocating the reserved channels to the backup paths.

We assume that provisioning would be based on distributed topology updating and signaling approaches, using protocols such as GMPLS [198, 241], and possibly proposed extensions to the signaling messages [199, 200]. Without loss of generality we assume that all the lightpaths are bidirectional. We also consider the case of opaque cross-connects only. All the OXCs in the network terminate the optical signal and switching, monitoring, and control and signal regeneration functions are performed in the electrical domain. Furthermore, wavelength conversion is available everywhere in the network and is not an issue here (as opposed to the case of all-optical switches). The case of transparent optical networks implies a different set of constraints [263, 264]. It requires protocols and algorithms that are different from the ones described below, and some of our claims regarding the performance of the considered failure recovery strategies may not apply to this type of networks.

The chapter is organized as follows. Section 9.2 analyzes the complexity of the deterministic approach to identify shareable backup channels, and Section 9.3 describes the details of the probabilistic approach. Section 9.4 describes an algorithm to compute shared backup path protection using the probabilistic approach. Since this approach does not provide the channels along the backup paths, Section 9.5 describes a distributed algorithm to optimize this channel assignment separately. Section 9.6 discusses the required extensions to the routing protocols, and Section 9.7 compares the performance results for realistic topologies using both probabilistic and deterministic-based shared backup path computation algorithms. We conclude in Section 9.8.

9.2   Complexity of the Deterministic Approach

In this section we compare the complexity, expressed in terms of processing time, and amount of information required by the algorithm, when determining a shared backup path in a failure dependent and a failure independent strategy. In both strategies we assume that the primary lightpath is given, and that a deterministic approach is employed. Note that we measure here the complexity of computing the backup path of a new service. This time should not be confounded with the failure recovery latency, which is the delay required to recover all the services on the precomputed backup paths when failures occur.

In the following, we denote by h the average primary path length expressed in number of links, and by h' the average length of the backup path (usually $h' \geq h$.) We use m to denote the number of links present in the network. We also assume that the total number of SRGs per reserved channel is of the order of $O(m)$. The total number of backup channels reserved throughout the network is denoted by x.

9.2.1   Complexity of the Failure Dependent Strategy

The failure dependent strategy only requires that sufficient backup channels be reserved on each link so that in any failure event all the affected backup paths can be accommodated. Suppose that every link maintains an associative array indicating for each SRG the number of times the SRG is traversed by a primary lightpath whose corresponding backup path traverses the link. The maximum value in this associative array is thus the maximum number of backup paths that would be concurrently activated on this link in a worst-case scenario. Therefore, if backup channels are assigned during failure recovery, a sufficient condition to guarantee full recovery of a single SRG failure is that each link of the network must have enough channels to accommodate its respective maximum number of concurrently activated backup paths. The latter condition can be determined in $O(mh)$-time based on the information available in the associative arrays. The combined size of the arrays is of the order of $O(m^2)$, which is reasonable for link-state dissemination protocols such as OSPF.

9.2.2   Complexity of the Failure Independent Strategy

In a failure independent strategy, the backup channels must be specifically assigned to the backup paths in a way that satisfies any foreseeable combination of backup path activation. In what follows, a list of SRGs protected by a given reserved channel consists of all distinct SRGs traversed by all the primary lightpaths whose respective backup paths are assigned the reserved channel. Thus a reserved backup channel can be reused to protect a primary path if no SRG traversed by the primary path appears in the list of SRGs already protected by the channel.

Shareable backup channels in the network are identified by verifying that for each backup channel in each link, the list of SRGs protected by the channel does not intersect with the list of SRGs traversed by the backup's primary path. Therefore, the complexity of identifying all the links with shareable backup channels in the network is $O(hx)$. This complexity assumes that each backup channel maintains a fixed length array in which each entry indicates whether an SRG is protected or not. It becomes $O(hx \log m)$ if instead a variable length list of protected SRGs is used. The number of backup channels is a function of g, the number of lightpaths in the network, and can be approximated by $x = O\left(gh'\right)$. Substituting $O\left(gh'\right)$ for x, the complexity of identifying the links with shareable channels is $O\left(ghh'\right)$. This operation requires that the list of SRGs protected by each backup channel be known. The size of this information is of the order of $O\left(gmh'\right)$. Our primary concern here is the dependence of this time and size complexity on the number of light-paths established in the network. We thus propose to substitute this time-consuming deterministic approach for a probabilistic approach whose complexity remains constant with respect to the number of lightpaths.

9.3   Probabilistic Approach

In what follows, we assume that the algorithm responsible for the route computation has an up-to-date knowledge of the state of the network that includes for each link (u, v) the number M of backup channels on link (u, v), and for every SRG j in the network the number n_j of backup channels on link (u, v) that protect a primary path traversing the SRG. Unknown is the identity of the backup channels that protect each SRG, or conversely the list of SRGs protected by each backup channel which was required in the case of the deterministic approach.

We now describe the technique used to quickly compute the probability that a backup channel is shareable with respect to a given primary path, using only the information specified above. We first present a combinatorial problem, and show that the solution to this problem provides us with the desired probability. Note that the computation of the exact probability as described in this section can be cumbersome. In practice we will in fact prefer an approximation, which is easier to compute and is sufficiently accurate for our purpose. The exact probability is thus provided here only as an exercise and as a means to gain a better understanding of the theory behind shared backup path protection.

9.3.1   A Problem of Combinations

The problem:

The problem of combinations is illustrated in Figure 9.2. We are given N bags numbered 1 to N, filled with colored marbles. Bag j ($j \in \{1, \ldots, N\}$) contains n_j marbles. All marbles in any given bag have the same color and are identical. Marbles coming from different bags are of different colors and are distinguishable. We are also given M identical buckets numbered 1 to M. We assume that the buckets have infinite capacity. We assign each marble to a bucket, selecting the buckets according to

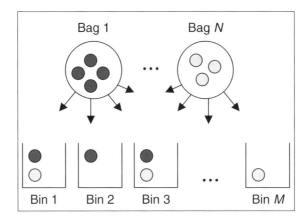

Figure 9.2: Bins and bags problem. (From [51], Figure 3. Reproduced by permission of © 2002 The Institute of Electrical and Electronics Engineers.)

a uniform random distribution with the condition that a bucket cannot contain two identical marbles. We want to compute:

1. How many differentiable combinations (denoted by Q) of marbles to buckets are possible?

2. Out of all combinations computed in (1), how many of them (denoted by D) have empty buckets left?

3. What is the probability that at least one bucket is empty?

The answers:

In the following we use $C(p, q) = q! / (p! (q - p)!)$ to denote the unordered combinations of p out of q elements.

1. First note that a solution exists if and only if $M \geqslant n_j$, $\forall j \in \{1, \ldots, N\}$. The marble arrangement of each bag into the M bucket is not conditional on other bags' arrangements. For each bag j there are $C(n_j, M)$ possible ways to arrange the n_j marbles into an ordered set of M buckets. There are thus $Q = \prod_{j=1}^{N} C(n_j, M)$ differentiable arrangements.

2. First note that if there exists $j \in \{1, \ldots, N\}$ such that $M = n_j$ then there is no combination that satisfies the constraint that a bucket cannot contain two marbles of the same color, and the answer is $D = 0$. Note also that if $M > \sum_{j=1}^{N} n_j$ then we cannot fill all the buckets, and the answer is $D = Q$. If $M = 1 + \max_{1 \leqslant j \leqslant N} n_j$, the problem is equivalent to question 1 posed in this section, except that now arrangements are confined to $M - 1$ buckets, the last bucket being left empty. There are thus $D = C(M - 1, M) \prod_{j=1}^{N} C(n_j, M - 1)$ possible arrangements. Otherwise, if $M > 1 + \max_{1 \leqslant j \leqslant N} n_j$, then $D = r(M - 1)$, where $r(k)$ is the recursion over integers in $k \in \{\max_{1 \leqslant j \leqslant N} n_j, \ldots, M - 1\}$ such that $r(\max_{1 \leqslant j \leqslant N} n_j - 1) = 0$, and $r(k) = \left[C(k, M) \prod_{j=1}^{N} C(n_j, k) \right] - r(k - 1)$ (see [52]).

3. Note that because there is no particular preference in how buckets are selected to receive a marble, the probability of occurrence is the same for all combinations computed in (1) and (2). The probability that at least one bucket is empty is thus equal to the ratio of the number of all the combinations with empty buckets to the total number of combinations, that is: $P = D/Q$.

Therefore, the exact probability that at least one bucket is empty is:

$$P = \frac{r(M-1)}{\prod_{j=1}^{N} C(n_j, M)} \tag{9.1}$$

$$r(k) = \left[C(k, M) \prod_{j=1}^{N} C(n_j, k) \right] - r(k-1)$$

$$r(\max_{1 \leqslant j \leqslant N} n_j - 1) = 0$$

$$k \in \{ \max_{1 \leqslant j \leqslant N} n_j, \ldots, M-1 \}$$

The estimated probability:

Note that the computation of D and Q may be tedious. We thus show here a means to approximate this probability. First observe that the probability that at least one bucket is empty is complement to the probability that all buckets are nonempty. And the probability that a bucket is nonempty is the complement to the probability p that this bucket is empty. Although p is conditional on the probability of other buckets being empty we assume that it is independent and identical for all buckets. Therefore, given a bucket, the probability p that the bucket is empty is the product of independent probabilities that all marbles of each bag are in other buckets, that is $p = \prod_{j=1}^{N} (1 - n_j/M)$. Based on our observations and assumptions, the probability that at least one bucket is empty is $P = 1 - (1 - p)^M = 1 - \left(1 - \prod_{j=1}^{N} (1 - n_j/M)\right)^M$. The complexity of computing P (or its complement $1 - P$) involves computing N products and an Mth power. It is realizable in $O(N + \log M) \approx O(N)$ time. The estimated probability that at least one bucket is empty is thus:

$$P = 1 - \left(1 - \prod_{j=1}^{N} \left(1 - \frac{n_j}{M}\right)\right)^M \tag{9.2}$$

9.3.2 Analogy with SRG Arrangement into a Set of Backup Channels

Assume that the M buckets of the problem presented in Section 9.3.1 are the backup channels on a given link. And assume that the N bags represent a list of N SRGs traversed by the primary path for which a backup channel is sought. The n_j marbles denote the number of times each SRG on the list is protected (through preestablished paths) by the backup channel set. Evidently, the same SRG cannot be protected multiple times by the same backup channel, otherwise contention would exist through their respective primaries if the SRG fails. This restriction is expressed in the problem formulation by the fact that two identical marbles (same SRG) cannot fall into the same bucket (backup channel). Thus, the problem above deals with computing the probability that there is at least one shareable backup channel, i.e. a backup channel that does not contain any of the N SRGs. We have shown that this probability is approximated in $O(N)$ time, where N is the number of SRGs on the primary path. Typically N is the average path length h. Therefore, the time complexity of identifying all the links with shareable backup channels in the network is $O(hm)$. This complexity is to be compared with $O(ghh')$ of the deterministic approach.

Remember that in the computation of these probabilities we have made two simplifying assumptions: (i) the probability of a backup channel being shareable is pairwise independent of other backup channels, and (ii) SRGs are uniformly distributed across backup channels. The effect of the first assumption can be quantified by way of simulations (see Section 9.7). The effect of the second assumption on the other hand is subtler because it depends on the policy used for allocating backup channels. For instance a *First Fit* or *Max Fit* policy tends to protect more SRGs with some backup channels than with others within the same link. As it turns out, a First Fit policy increases the probability that a backup channel is available compared to an allocation that picks the backup channels at random according to a uniform distribution.

9.4   Probabilistic Routing Algorithm with Partial Information

We describe here in detail an algorithm that implements the probabilistic approach, and compare it with the equivalent deterministic algorithm.

Given: A topology represented as a graph $G(V, E)$ where vertices represent optical cross-connects (OXC) and links represent fiber strands between OXCs. A network state database, that indicates for each link the number of channels available, the number of backup channels protecting preestablished lightpaths and the number of times each SRG in the network is protected by a backup channel in that link. The latter information is stored into an array. The array's indices correspond to SRGs and each entry in the array counts the number of backup channel cross-connections that would occur in the link if the corresponding SRG fails.

Input: A pair of nodes $A-Z$.

Output: A pair of bidirectional lightpaths from A to Z, primary and backup with minimum cost, excluding backup channels that are shared with preestablished backup paths.

The algorithm *ComputeRoute* is illustrated in Table 9.1.

Table 9.1: The ComputeRoute pseudo-code

1. Compute K-shortest paths. Sort the paths by length and label them w_1 to w_K.
2. Set $S = \emptyset$.
3. For each shortest path w_i:
(a) To each link that shares an SRG with w_i or has neither available channel nor backup channel, assign infinite weight.
(b) For each link without a backup channel, set weight to cost of link.
(c) For each link with backup channels, set weight to cost of link times the probability that no backup channel is shareable (by way of the approach presented earlier in Section 9.3, Equation 9.2).
(d) Compute the shortest path s_i using the metric defined in steps 3a to 3c, and set $S \leftarrow S + \{w_i, s_i\}$.
4. Select the minimum cost path pair $\{w_K, s_K\} \in S$.
5. If no path can be found in step 4, return NO_PATH, otherwise return $\{w_K, s_K\}$.

Upon a request to provision a new service, the ComputeRoute procedure is invoked to compute a *probably* most cost-efficient route. This route is not guaranteed to be feasible, however, because of its probabilistic nature, or because the information used to compute the route could be outdated. Feasibility is thus verified during path setup, when cross-connections are created along the path. Backup channels are assigned locally, using a first-fit approach, or using some local optimization algorithm (as described later in Section 9.5). If the local channel assignment procedure cannot find a shareable backup channel, it creates one from the pool of available channels or fails if the pool of available channels is exhausted. If path setup fails, the signaling clears the partially created cross-connections, and returns an error message identifying the set of links that are unable to satisfy the request. The returned links are removed from the network, and a crankback mechanism invokes the ComputeRoute procedure on the reduced network to compute an alternate route. This iteration is repeated up to a maximum number of crankbacks before giving up, or until a route is successfully set up. In Section 9.7, we present results that assume that backup channels are assigned on a first-fit basis during path setup signaling, and are reoptimized at the end of the simulation using the local channel assignment optimization described later in Section 9.5. We make this assumption only for experimental purposes; in a distributed environment one could use the channel assignment optimization during path setup. We also assume that channels are always available (uncapacitated case) and therefore crankback is not required. This is a realistic assumption since network deployment activities are usually planned to maintain network utilization below 70%.

The algorithm is self-explanatory. It differs from the deterministic algorithm only in step 3(c) of Table 9.1. In the deterministic algorithm the weight of a link is set to the link cost times ϵ ($\epsilon \ll 1$) if it contains a shareable backup channel and to the link cost if it does not (see Chapter 7). In the probabilistic algorithm this weight is replaced by the cost of the link times the probability that no backup channel is shareable in the link. Note that the deterministic approach requires additional information to compute the routes. In particular, it needs to know whether each SRG is protected or not for every backup channel. This is contrary to the probabilistic approach, where only the number of times an SRG is protected in every link by any backup channel of that link needs to be known. Finally, note that we separated lightpath provisioning from routing, and channel assignment is performed in a distributed way after the lightpaths are selected by the route computation. The objective of the route computation is to determine the paths so that sharing is maximized during channel assignment. Even though a link may be erroneously tagged as having a shareable channel during path computation, the channel assignment procedure during path setup will guarantee that there are no sharing violations. In order to guarantee this, the scheme used for channel assignment requires the same information as for the deterministic approach. However, we will see that this information can be distributed across the nodes in the network. It suffices that each node maintains a local database containing only information related to the backup channels terminating into it. We address this backup channel assignment problem in the next section.

9.5 Locally Optimized Channel Selection

9.5.1 Shared Mesh Protection Provisioning Using Vertex Coloring

Recall that two shared backup paths are *compatible* and may share a backup channel if their respective primary paths are SRG-disjoint. Otherwise they are said to be *conflicting*. Although only single SRG failures are considered here, the description of the algorithm can easily be transposed to protect against node failure as well: replace SRG by node where it applies. Given a group of protection paths traversing a common link, the problem is to assign the minimum number of backup channels to the paths in the link in accordance with the rules of sharing. Typical online provisioning algorithms assign backup channels on a first-come first-serve basis and reserve new channels when sharing is not possible with present backup channels. In this approach the number of backup channels depends ultimately on the order of arrival of the backup paths. Since the order cannot be determined in advance, an optimization algorithm must be invoked at regular intervals to reassign the channels to compensate for suboptimality. In this section, we show that finding the optimum assignment is equivalent to solving a vertex-coloring problem.

The allocation of shared backup channels is tantamount to a vertex-coloring problem: given the set of all backup paths that intersect on a given link, represent every path as a vertex, and connect with an edge every pair of vertices whose corresponding paths are conflicting. Assign a distinctive color to each backup channel, and allot a backup channel to each path, that is color the vertices. Clearly, two vertices cannot be allotted the same color if they are connected by an edge, since the corresponding backup paths are conflicting and cannot share a channel. The objective is to minimize the number of backup channels (number of colors) required to accommodate all backup paths (color all vertices), while avoiding conflicts.

This problem is known to be NP-hard, however there are many heuristics that can be used to compute suboptimal solutions. A vertex-coloring algorithm that offers a good trade-off between quality and runtime complexity is DSATUR [59].

9.5.1.1 Example

Consider the example of Figure 9.3. The figure illustrates five lightpaths $\{AD, CD, BC, AC, BD\}$ and their backup paths, routed in a four-node ring network. All the backup paths traverse link $C-D$. The demands are provisioned following the sequence indicated in the table in Figure 9.3(b). If we use a typical online shared backup path provisioning with a first-fit channel assignment strategy, and apply the graph representation presented earlier to $C-D$, we obtain the coloring shown in Figure 9.3(c). Even though a single failure in this example affects at most three primaries, this coloring consumes four colors, indicating that four backup channels are required. An optimized coloring yields the solution shown in Figure 9.3(d), which consumes only three colors. Comparing Figures 9.3(c) and 9.3(d), we observe that a new channel (R) should have been allotted to the backup path of demand (BC) instead of sharing channel (B) with the backup path of demand (AD). This solution, however, is not considered because it is not optimal when the third demand is being provisioned (that is demands $\{AD, CD\}$ are routed and request for demand (BC) has not yet arrived) since at that time it would require three channels (B, G, R) instead of two (B, G).

9.5.2 Implementation and Applications

In the probabilistic routing algorithm the backup channels are not determined by the routing algorithm. Therefore, they need to be determined by the optimized channel assignment procedure each time a lightpath is being provisioned and signaled. Furthermore, the optimized channel reassignment can be a low priority process running in the background upon certain events, or at regular intervals. The information necessary to accomplish this task is available locally in every OXC and is independent of nonadjacent OXCs. Thus, each OXC can run a copy of the algorithm in a distributed manner, locally and independently of other OXCs. A change in the allocation of a backup channel need only be propagated to its endpoints. Since backup channels are only reserved and are not cross-connected until a failure recovery occurs, the task amounts to no more than modifying and exchanging local channel-sharing information between pairs of adjacent nodes. For every OXC-pair connected by at least one channel, the OXC with the highest IP address is delegated to perform the task.

A byproduct of the optimized channel reassignment is that it can also be used to migrate the backup paths of dedicated backup path protection to shared backup path protection if desired. By changing their failure recovery type to shared backup path protection, we allow the background optimization process to apply the channel reassignment optimization to these services. Note that the algorithm does not optimize the routes of the backup paths, and the resulting solution is not nearly as efficient as a

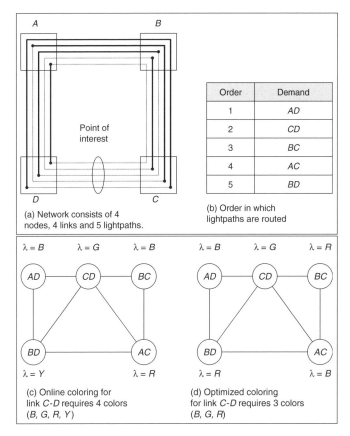

Figure 9.3: Example of sub-optimal (first-fit) and optimal local backup channel assignment. (From [52], Figure 7. Reproduced by permission of © 2004 The Institute of Electrical and Electronics Engineers.)

reoptimization algorithm that reroutes the backup paths to maximize sharing [55]. Such an algorithm is outside the scope of this chapter and will discussed later in Chapter 10.

Finally, the channel reoptimization procedure closes an advantage gap of the failure dependent strategy over the failure independent strategy in the case of multiple failure scenarios. Since the backup channels are preassigned in the failure independent case, there is a higher probability that two services affected by two distinct failures contend for the same protection backup, even if there are backup channels available on the same span [17, 104]. Reprovisioning mechanisms that compute backup paths dynamically when the planned failure recovery fails would mitigate this problem, but are not covered here [251]. Then again, a background channel reoptimization process would detect the prospect for such contentions after the first failure, and reassign the channels to eliminate them as they occur.

9.6 Required Extensions to Routing Protocols

As discussed earlier, for distributed path computation, scalability issues arise for large networks in terms of required control network bandwidth and memory if the complete link resource availability and sharing information are distributed to every node. This is mitigated in the case of the probabilistic approach. The local database of each node contains a summarized information that is necessary to compute the routes using the probabilistic approach. For optical cross-connects with hundreds of ports there may be multiple data links between a pair of nodes. In order to improve scalability, data links between the same pair of nodes, with similar characteristics, can be bundled together and advertised as a single link bundle or a traffic engineering (TE) link into the routing protocol [38, 219]. The path computation algorithm requires the network topology and optical link resource information. It is the responsibility of the routing protocols to disseminate this information. OSPF with traffic engineering and GMPLS extensions is able to disseminate sufficient information for computing unprotected and dedicated backup paths. However, to compute the backup path for shared backup path protection, the path computation module needs to have the resource-sharing information of the links in the network, as discussed in Section 9.3. To support path computation for shared backup path protection, all or some of the information below should be disseminated by the routing protocol, depending on the path computation algorithms:

- Summarized information about the backup resource sharing on a TE link for shared backup path protection, including the total number of backup paths sharing the channels reserved on the TE link for shared backup path protection, the total number of SRGs protected by the backup channels on the TE link, and the total number of shared backup channels.

- The list of SRGs protected by the backup channels on the TE link and their respective shared backup channels since SRG-disjointness is required to guarantee recovery in the event of a single SRG failure.

The list of SRGs protected by the TE link is defined as the union of SRGs traversed by all the primary paths whose respective backup paths share the reserved backup resource on this TE link. The shared backup bandwidth for an SRG indicates the available backup channels on the TE link that can be reserved for recovering from this SRG failure. If a primary path only traverses one SRG, the available backup bandwidth that its backup path can share on this TE link is the shared backup bandwidth for this SRG. When a primary path traverses multiple SRGs, the shared backup bandwidth available for its backup path is the intersection of the shared backup bandwidth of the individual SRGs, and becomes smaller as the number of SRGs increases. The total shared backup bandwidth is the bandwidth reserved on the TE link for failure recovery, which is the union of the shared backup bandwidth for all SRGs and nodes.

OSPF traffic engineering extensions [175] and GMPLS extensions [181] make use of the Opaque LSA (Link State Advertisement). An Opaque LSA, called Traffic Engineering LSA, carries the additional attributes related to TE and GMPLS links, and standard link-state database flooding mechanisms are used for distributing TE LSAs. The LSA payload consists of one or more nested TLV (Type, Length, Value) triplets for extensibility. There are two types of TE LSAs [84]. One contains a Router Address TLV (Router Address TE LSA) that specifies a stable IP address of the advertising node. The other contains a link TLV (Link TE LSA), which describes a TE link. The link TLV is constructed of a set of sub-TLVs that specify the link attributes. We only consider the Link TE LSA here, and refer to it as TE LSA unless otherwise stated. Each TE LSA carries the summarized resource information for a TE link. For each TE link, the attributes, including link type, TE metric, available resource, administrative group, local and remote link identifier (or interface IP addresses), link protection type, and interface switching capability descriptor, are specified in the form of sub-TLVs in the link TLV of the TE LSA. The information in the TE LSAs is used to build an extended TE link-state database for the explicit path computation, just as router LSAs are used to build a regular link-state database for packet forwarding.

The extensions in support of carrying link-state information for the path computation of shared backup path protection can be based upon the OSPF-TE and its GMPLS extensions. Specifically, the sub-TLVs carrying the above sharing information of the backup resource on a TE link can be added to the link TLV of the TE LSA so that the information can be used by the path computation algorithm to compute the backup path. Two sub-TLVs can be defined [199]. The failure recovery information summary sub-TLV specifies the sharing information of the backup resource reserved for the shared backup path protection on the TE link, including number of SRGs protected and total shared backup bandwidth (i.e. number of reserved channels). The SRG shared backup bandwidth sub-TLV identifies the shared backup bandwidth for a protected SRG on the TE link.

9.7 Experiments and Performance Results

9.7.1 Accuracy and Distributions of Probability Functions

For the experiments presented in this section we use both the deterministic and probabilistic implementations of the algorithm. Great care must be taken in optimizing the deterministic implementation for speed. The probabilistic code can then be derived from the deterministic code by modifying step 3(c) as described in Table 9.1. We use here the same first-fit channel assignment algorithm in both implementations in order to isolate and limit our measurements to the effects of using a probabilistic approach. The benefits of a local channel assignment optimization are measured separately in Section 9.7.3.

In the following we first measure the quality of the estimated probability that a link contains a shareable backup channel given the information on the number of times each SRG traversed by the primary path is recovered on that link. The experiment consists of simulating a large number of arbitrary instances of the problem presented in Section 9.3.1. For each instance of the problem, we use a Monte-Carlo method [119] to generate several millions of randomly selected arrangements, and compute the ratio of generated combinations with available backup channels to the total number of generated combinations (i.e. estimate Equations 9.1 and 9.2 computed in Section 9.3.1). We then compare the difference between each experimental probability and the corresponding exact and approximate probabilities obtained by computation. The results are shown in Figures 9.4 and 9.5. Figure 9.4 demonstrates the error distribution of the exact probabilities minus experimental probabilities obtained over the range of problem instances. The simulation exhibits an accuracy within 0.01 of the exact probability, and a closer look even indicates that 70% of the time the difference is within 5×10^{-4}. In comparison, we observe in Figure 9.5 that the estimate probability has a tendency to underestimate the experimental probability, but it is accurate within 0.05 for 85% of the time, which is quite remarkable given the simplicity of the computation. Note that this difference was expected due to the independence assumption made in the determination of Equation 9.2.

9.7.2 Comparison of Deterministic vs Probabilistic Weight Functions on Real Networks

In the next set of experiments we consider two scenarios inspired from real-life networks. NetA is a 100-node, 137-link network, with one unit of demand between every pair of nodes (4950 demands). NetB is a 220-node, 300-link network, also with one unit of demand between every pair of nodes (24 090 demands). For the sake of simplicity we assume here that every link costs one unit of currency and corresponds to one SRG (i.e. one SRG per link and one link per SRG). We also assume that capacity is abundant, and path setup always succeeds without requiring crankback. We then route the demands on each network using the deterministic and the probabilistic algorithms. We are interested here in the processing time to complete each algorithm, and the quality of the solutions expressed in total number of channels required (used for primaries and reserved for backups). Tables 9.2 and 9.3 summarize the results. For NetA (respectively NetB) we observe that the probabilistic approach is 6.78 times faster (respectively 19.7 times faster) than the deterministic approach while the capacity penalty is only 2% (respectively 3%). Also important is the amount of information the route computation algorithm requires in order to find the routes. The probabilistic-based route computation procedure

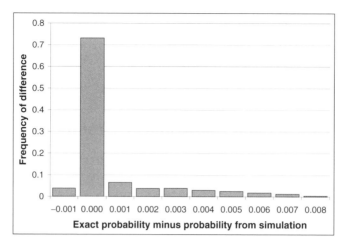

Figure 9.4: Error distribution of exact sharing probability minus probability from simulation. (From [51], Figure 5. Reproduced by permission of © 2002 The Institute of Electrical and Electronics Engineers.)

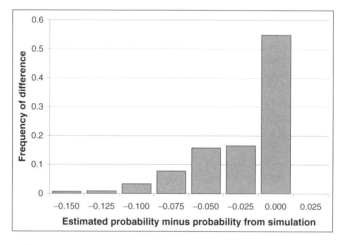

Figure 9.5: Error distribution of estimated sharing probability minus probability from simulation. (From [51], Figure 4. Reproduced by permission of © 2002 The Institute of Electrical and Electronics Engineers.)

only requires one array per link, where each entry in the array indicates the number of times the SRG is protected on the link by any backup channel. For instance, in the NetB problem, there are 300 such arrays (one per link) of 300 entries each (one per SRG). For comparison, the deterministic approach needs one array for each backup channel, where each entry in the array corresponds to an SRG and indicates whether the SRG is protected or not by the backup channel. In the solution of the NetB problems, 213 052 of the channels are reserved for protection; thus 213 052 arrays of 300 entries would be required in the deterministic method.

For the same set of experiments, Figure 9.6 plots the distribution of sharing probabilities as computed in step 3(c) of the probabilistic algorithm (Table 9.1) during the provisioning of the demands in

Table 9.2: Time required for the deterministic and probabilistic algorithms to complete (in seconds). (From [51], Table 1. Reproduced by permission of © 2002 The Institute of Electrical and Electronics Engineers.)

Network	Deterministic	Probabilistic	Ratio
NetA	156	23	6.78:1
NetB	9885	501	19.7:1

Table 9.3: Capacity usage for the deterministic and probabilistic algorithms (as total number of channels). (From [51], Table 1. Reproduced by permission of © 2002 The Institute of Electrical and Electronics Engineers.)

Network	Deterministic	Probabilistic	Ratio
NetA	61312	62716	100:102
NetB	520771	536343	100:103

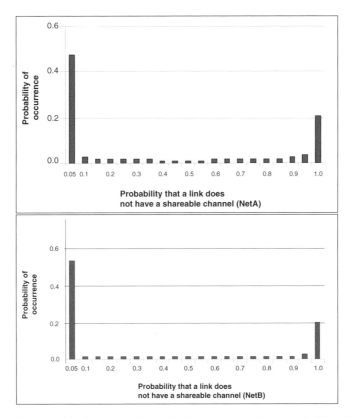

Figure 9.6: Distribution of sharing probabilities (NetA, top and NetB, bottom). (From [51], Figure 6. Reproduced by permission of © 2002 The Institute of Electrical and Electronics Engineers.)

Table 9.4: Comparison of total number of channels required for deterministic, probabilistic and distributed (shortest disjoint paths) approaches. Percentages are relative to deterministic approach. (From [52], Table 3. Reproduced by permission of © 2004 The Institute of Electrical and Electronics Engineers.)

Network	Deterministic	Probabilistic	Shortest disjoint paths
NetC	3884	3908 (100.6%)	4217 (108.6%)
NetD	897	901 (100.5%)	1055 (117.5%)
NetE	1194	1237 (103.6%)	1391 (116.5%)

NetA and NetB. The distributions for the two networks are similar and show that 70% of the time in NetA (77% in NetB) it was possible to predict almost certainly whether there would be a shareable backup channel (probability 0.0 that a link does not have a shareable channel, 48% of the instances for NetA, and 57% of the instances for NetB) or not (probability 1.0, 22% of the instances for NetA, and 20% of the instances for NetB).

The last experiment of this section compares the deterministic and the probabilistic-based algorithms with a distributed algorithm that was first described in [251]. This third algorithm computes a pair of disjoint paths based on the topological information without trying to share existing backup channels, for which we assume that the information is not globally available. As for the probabilistic approach, the assignment of the backup channels is done locally during path setup signaling when the paths are provisioned. For this comparison we test the three algorithms, deterministic, probabilistic and shortest disjoint paths, on three real carrier networks with realistic demands, NetC, NetD and NetE. NetC is a 17-node, 26-link network, NetD is a 45-node, 77-link network, and NetE is a 50-node, 88-link network. The results, presented in Table 9.4, indicate that the probabilistic approach is comparable to the deterministic approach. In comparison, the third approach, which ignores the possibility of sharing existing backup channels, performs relatively poorly, and requires from 8% to 17% more channels than the other two algorithms.

9.7.3 Benefits of Locally Optimized Lightpath Provisioning

In the next set of experiments we compare the benefits of local backup channel optimization on two realistic core mesh networks. This procedure is independent of the method used to compute the paths, and we thus use it in combination with the deterministic routing algorithm only. Network A consists of 46 nodes interconnected by 75 links and loaded with 570 lightpaths. Network B consists of 61 nodes, 88 links, and 419 lightpaths. For each network, we provision all the demands in sequence using various values of demand churns. The demand churn is the amount of demand expressed as a percentage of the total routed demand, which after some time is taken out of service and removed from the network to leave room for subsequent demands. The rate at which demands are removed is determined such that if the churn is C, then at the end of the simulation the network contains $(100 - C)\%$ of the total demand, and the remaining $C\%$ will have been routed and then removed before the end of the simulation. We use a first-fit backup channel assignment during provisioning, and apply a local backup channel optimization after all the demands are routed. We measure the number of backup channels required before and after local channel assignment optimization and report the savings as a percentage of total backup capacity in Figure 9.7. Our measurements indicate that as the demand churn increases, the number of backup channels that can be freed becomes substantial.

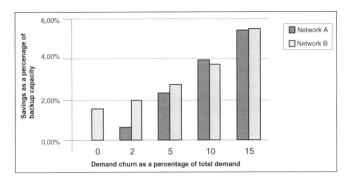

Figure 9.7: Savings generated by local optimization as a percentage of backup capacity, as a function of demand churn. (From [52], Figure 11. Reproduced by permission of © 2004 The Institute of Electrical and Electronics Engineers.)

9.7.4 Summary

Before we conclude, we summarize the results section with a comparison of the different approaches presented earlier (Table 9.5). Four strategies are compared, each corresponding to a column in the table. They are the failure dependent strategy using channel pooling, the centralized failure independent strategy, without and with local reoptimization of the backup channels, and the distributed failure independent strategy with local reoptimization of the backup channels. We evaluate the strategies according to six performance factors, one for each row of the table.

9.8 Conclusion

In this chapter we have addressed some of the scalability issues that are most likely to be encountered when designing and operating shared backup path protection architectures. We have seen in particular that one of the challenges is to identify deterministically the existence of backup channels allocated to pre-existing demands that can be reused (i.e. shared) when computing the backup paths for a new demand. Using a probabilistic approach, we show that summarized information consisting of one fixed length array for every link is sufficient to compute the paths efficiently while maximizing sharing opportunities. In contrast, the deterministic approach needs one such array for every backup channel, and thus does not scale when the demand grows and the number of backup channels increases.

Our experimental results demonstrate that the probabilistic approach completes the routing $6-20$ times faster than the deterministic approach for networks ranging from 100 to 200 nodes. Although the probabilistic approach uses several orders of magnitude less information than what is necessary for a deterministic approach, their solutions are within $2-3\%$ of each other in terms of capacity usage. In fact our experiments indicate that 70% of the time this limited information is sufficient to determine with certainty whether or not there exists a protection channel on a link that could be shared.

One possible and natural application of the probabilistic approach is the distributed routing of SBPP lightpaths utilizing optical switches. The local database of each switch may contain summarized information that is necessary to compute the routes using the probabilistic approach. Since this information is limited, it can be easily disseminated by link-state protocols, such as OSPF. Using this

Table 9.5: Comparison of the route-provisioning algorithms presented. (From [52], Table 4. Reproduced by permission of © 2004 The Institute of Electrical and Electronics Engineers.)

	Failure dependent strategy using channel pooling (centralized or distributed)	*Centralized, failure independent strategy with no local reoptimization of protection channels*	*Centralized, failure independent strategy with local reoptimization of protection channels*	*Distributed, failure independent strategy, with local reoptimization of protection channels*
Time complexity of provisioning algorithm	$O(hm)$ Fast.	$O(ghh')$ Slow, but speed is not an issue for provisioning purposes.		$O(hm)$ Fast.
Amount of information required by provisioning algorithm	$O(m^2)$ Minimal.	$O(gh'm)$ Large. Grows with traffic.		$O(m^2)$ Minimal. Well adapted for link-state advertisement protocols such as OSPF [199, 235].
Protection capacity required	Most capacity efficient of all presented strategies.	Slightly less capacity efficient than failure independent strategy with local reoptimization of protection channels.	Nearly as capacity efficient as failure dependent strategy.	Slightly less capacity efficient than centralized failure independent strategies.
Recovery latency	Slow. Requires inter-node communication to agree on a protection channel assignment during failure.	Fast. Each node knows immediately which protection channels to cross-connect, because this information is predetermined and available locally.		
Availability in case of multiple failures	High availability. Protection channels can be assigned to accommodate multiple failures.	Prone to protection channel contentions in case of multiple failures. This can be mitigated by re-provisioning mechanisms [17].	At least as good as failure dependent strategy (protection channels can be locally reoptimized after first failure).	
Complexity of implementation	Very complex. Requires communication between adjacent nodes to agree on selected channels and to remove channel selection conflicts.	Complex.	Complex.	Complex.

information, for each demand, the ingress switch can compute a path equivalent to a path computed by a centralized deterministic algorithm with a complete view of the state of the network.

We also discussed a distributed method that rearranges the allocation of shared channels reserved for recovery, with the objective of minimizing the number of allotted channels. This algorithm can be implemented as an independent background process to supplement either the centralized or distributed provisioning algorithms. It is effective to correct suboptimality inherent to a first-fit based provisioning, or seize on improvement opportunities that are brought forth by demand churn.

Chapter 10

Path Reoptimization

10.1 Introduction

It is now well understood that the intelligent mesh optical networks deployed today offer unparalleled capacity, flexibility, availability, and, inevitably, new challenges to master all these qualities in the most efficient and practical manner. More specifically, requests for services are received and routed using an online routing algorithm that takes all of the information available at the time of the request to make the appropriate routing decision. The primary and backup paths of each new demand are computed according to the current state of the network, which includes the routing of the existing demands. As the network and traffic evolve, the routing of the existing demands becomes suboptimal. Demand churn and network changes such as the addition or deletion of new links and or capacity, cause the routing to become suboptimal, thereby creating opportunities for improvements in network bandwidth efficiency. Increasing customer churn and the continued demand for bandwidth services exacerbate this problem.

Reoptimization seizes on these opportunities and offers the network operator the ability to better adapt to the dynamics of the network. This is achieved by regularly (or upon a particular event) rerouting the existing demands, temporarily eliminating the drift between the current solution and the best known solution that is achievable under the same conditions, as illustrated in Figure 10.1.

Carriers have been using reconfiguration over time to better manage their network assets and increase utilization of those assets, thereby deferring capital spending on new infrastructure. Reconfiguration has also been used to provide better service performance, for example by rerouting services over shortest paths if such paths become available. Earlier work on reconfiguration was done in the context of Digital Cross-Connect (DCS)-based networks [20, 25]. Later, with the deployment of ATM as a networking technology, reconfiguration of ATM networks was explored, leveraging ATM Virtual Paths [28, 151, 215, 227, 322]. Reconfiguration has also been studied in the general context of optical networks [31, 32, 44, 129, 191, 202, 261, 285, 279]. More recently, [196] discusses various strategies to maximize the benefit of span protection. In [139] the authors address the value of rearrangeability in incremental capacity planning for span-protected networks.

This chapter explores the benefits of reconfiguration, or reoptimization, of optical mesh networks, specifically those that offer SBPP services as described in earlier chapters. In the optical mesh network of interest here, carriers can either reroute only the shared backup paths of existing lightpaths, so that service is not affected (it is still carried on the primary or working path during the reoptimization of the backup paths), or they can reroute both the primary and shared backup paths (either by impacting

Path Routing in Mesh Optical Networks   Eric Bouillet, Georgios Ellinas,
Jean-François Labourdette, Ramu Ramamurthy   © 2007 John Wiley & Sons, Ltd

on customer service, or by first moving service to the backup path and reoptimizing the primary, and then moving service back to the primary and reoptimizing the backup). Reoptimizing both primary and backup paths should improve network bandwidth utilization more than just reoptimizing backup paths, but at the cost of customer service impact, or additional operational complexities and risks.

In this chapter we study two reoptimization algorithms. A complete reoptimization algorithm that reroutes both primary and backup paths, and a backup reoptimization algorithm that reroutes the backup paths only. Rerouting backup paths only is a suboptimal but attractive alternative that avoids any service interruption since the primary path is not affected. We show that on average, with periodic reoptimization, these algorithms allow bandwidth savings of 3–5% of the total capacity in scenarios where the backup path only is rerouted. Substantially larger bandwidth savings can be achieved when both the working and backup paths are rerouted. In addition, significant bandwidth savings can be achieved by reoptimizing the network after topological changes such as new nodes and/or new link additions. These bandwidth savings are achieved through increased sharing of backup path capacity among several working paths, and substantial reductions in average path length, which also translates into shorter recovery times. We also demonstrate that trying all possible demand permutations with an online algorithm does not guarantee optimality, and in certain cases does not achieve it, while for the same scenario optimality is achieved through reoptimization. This observation motivates the need for a reoptimization approach that does not just simply look at different sequences. We describe such an approach in this chapter, and experiment with it. Reoptimization has actually been performed in a nationwide live optical mesh network and we share the results of this experience in this chapter, validating the reality and the usefulness of reoptimization in real networks.

The chapter is organized as follows. In Section 10.2 we discuss the online algorithm cost model and the main function used to compute the shared backup paths that achieve the desired compromise between cost and recovery latency. In Section 10.3, we describe the reoptimization algorithm. This algorithm uses the routing function discussed in Section 10.2. Section 10.4 is a collection of proofs where we demonstrate the existence of cases for which neither an algorithm that tries different sequences to route the demands nor the proposed reoptimization algorithm can achieve the optimum solution. We also reveal the existence of cases for which reoptimization achieves the optimum, whereas trying all possible sequences to route the demands does not. Finally, the effectiveness of the reoptimization algorithm is measured for real customer networks and the results are presented in Section 10.5. Concluding remarks are offered in Section 10.6.

10.2 Routing Algorithm

10.2.1 Cost model

As in earlier chapters, we use the term Shared Risk Group (SRG) to indicate a group of optical resources that share a common risk of failure. For the reoptimization we use the cost model previously described in Section 7.2, which is briefly reiterated here. We define the *length* of a path as the sum of the predefined weights of the links (or channels) that constitute it. The metric or policy used for weighting the links is different for primary and backup paths. For primary paths it is the real cost c_e of using the links. This cost takes into account the price of the optics and the transponders, as well as the price of the common equipment such as optical amplifiers and WDM systems, which is equally distributed over the multiple channels they support. For a backup path it is a function of its primary path. A backup link e is assigned: (1) infinite weight if it intersects with an SRG of its primary path; (2) weight $w_e = c_e$ if new capacity is required to route the path; and (3) weight $s_e \leq w_e$ if the path can share existing capacity reserved for preestablished backup paths. The cost of a primary path and its protection path is then the sum of their respective lengths. Quite evidently, the underlying idea here is to encourage sharing, whereby existing capacity can be reused for routing multiple backup paths. The condition for sharing is that the backup paths must not be activated simultaneously, or

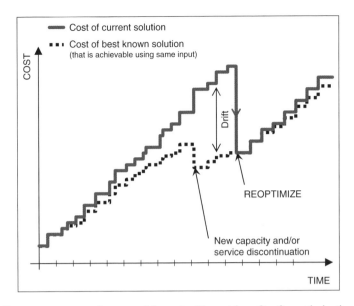

Figure 10.1: Current cost versus best possible cost with cost-benefit of reoptimization. (From [57], Figure 2. Reproduced by permission of © 2002 The NOC.)

in other words that their respective primaries must be pairwise SRG-disjoint so that they do not fail simultaneously. We call ϵ the ratio s_e to w_e (or c_e). This ratio can be adjusted for the desired level of sharing. For smaller values of s_e, and thus ϵ, backup paths will be selected with the minimization of the number of nonshareable links (weights w_e) in view, eventually leading to arbitrarily long paths (as expressed in number of hops) that consist uniquely of shareable links (weights s_e). For larger values of s_e, routing is performed regardless of sharing opportunities and backup paths will end up requiring substantially more capacity.

10.2.2 Online Routing Algorithm

We now describe a K-shortest path based algorithm that computes a primary and backup pair of routes, and assigns channels along the routes using the cost model described in Section 10.2.1. The algorithm is derived from the algorithm presented in Table 7.2, with the main difference that a primary path can be specified for backup reoptimization. It takes as input: (1) a network instance N that encapsulates the state information of the switches, optical channels (busy and available), and existing demands with their routes; (2) the end-nodes A and Z of the demand; and (3) a candidate primary path p_0 if backup reoptimization is desired. It operates as shown in Table 10.1.

If the minimum cost is sought (maximum sharing), the value of $\epsilon = s_e/c_e$ in step 5a of Table 10.1, determining the cost of shareable protection channels, is set to 0. Otherwise if shorter backup lengths and faster recovery times are desired, ϵ is set to a small positive value. Extensive studies have already been performed for $\epsilon = 0$ in [107], and in Chapter 7 we investigated the effect of varying ϵ between 0 and 1. When the value of ϵ moves towards 1, we expect the lengths of primary and backup paths, as expressed in number of hops, to resemble that of DBPP, though sharing is still implemented when available on the backup path and the capacity required remains lower than that required for DBPP. Earlier experiments ([55, 241]) indicate that a value of ϵ in the range $[0.2 - 0.4]$ returns the best trade-offs between cost and recovery latency (see Chapter 7). In the remainder of this chapter we use $\epsilon = 0.3$.

Table 10.1: Online routing algorithm

Compute_Pair_of_Paths(N, A, Z, p_0):

N the network

A an end-node

Z the other end-node

p_0 a candidate primary path if backup reoptimization is desired.

1. If p_0 is nonnull, set $P = \{p_0\}$ and go to step 4, otherwise compute a list of candidate primary paths.

2. For every link e in N set weight to cost c_e of one channel in link (cost of transponders, regenerators and OAs).

3. Compute set P of k minimum-weight paths connecting node-pair $A-Z$, or all feasible paths if there are less than k of them.

4. Set min_weight $=$ infinity, and $\{p^*, q^*\}$=INFEASIBLE.

5. For each primary path p in P:

   (a) Assign weight to every link e:

       i. If e intersects SRG of primary path p, set weight to infinity.

       ii. If e has at least one channel that is shareable with p, set weight to $s_e = \epsilon c_e$.

       iii. Otherwise, set weight to $w_e = c_e$.

   (b) Using metric defined in step 5a, compute minimum-weight backup path q connecting node-pair $A-Z$.

   (c) If q does not exist, continue step 5 with next path p in P.

   (d) If min_weight $<$ combined weight of paths p and q, then $\{p^*, q^*\} = \{p, q\}$ and min_weight $=$ combined weight of paths p and q.

6. Return $\{p^*, q^*\}$.

10.3   Reoptimization Algorithm

The reoptimization algorithm takes as input: (1) a network instance N that encapsulates the state information of the switches, optical channels (busy and available), and existing demands with their routes; and (2) a list D of demands to be reoptimized with their respective reoptimization types (*complete* reoptimization of working and backup paths, or reoptimization of backup paths only). It operates as shown in Table 10.2.

The reoptimization algorithm described in Table 10.2 consists of successive step improvements. This method has been used to solve a variety of problems in the field of network optimizations, because it is relatively simple to implement and it provides very good results. A similar approach was proposed in the case where protection channels are not preassigned [201]. In [49] the authors improve on the approach with a genetic algorithm that first generates permutations of the demand set D before applying the sequential or the reoptimization algorithm, and selects permutations that result in the

Table 10.2: Reoptimization algorithm

Reoptimize_Demands (N, D)

    N the network

    D the list of demands with respective reoptimization types

1. Set *REPEAT = FALSE*.

2. For each demand d in D:

    (a) Let A and Z denote the endpoints of the demand.

    (b) Set p_0 = current primary path and q_0 = current backup path of demand d.

    (c) In network N, free paths p_0 and q_0.

    (d) If backup reoptimization is desired:

        $\{p^*, q^*\}$ = Compute_Pair_of_Paths(N, A, Z, p_0),

        (note that $p^* = p_0$)

    else:

        $\{p^*, q^*\}$ = Compute_Pair_of_Paths(N, A, Z, *null*)

    (e) If combined weights of p^* and q^* are less than combined weights of p_0 and q_0, then in network N, route demand d on paths p^* and q^*, and set *REPEAT = TRUE*. Otherwise, route demand d back to paths p_0 and q_0.

3. If *REPEAT = TRUE*, repeat from step 1.

---

minimum cost. We prove in the next section that there are instances for which the reoptimization algorithm achieves the optimum routing configuration, while a sequential algorithm that tries to find a best ordering for routing the demands for the same instances (such as in [65] or [49]) would fail to find the optimum. We also prove that there are other instances for which both algorithms fail to find the optimum routing configuration even if all possible permutations of the demand set are explored. The reoptimization algorithm is generic enough so that it is also applicable to reoptimize mixed protection types, i.e. combination of unprotected, DBPP and SBPP demands of various rates. It is also fast and easy to enhance with additional rules that improve the quality of the reoptimization. It can for instance be improved to selectively reoptimize a specified set of demands. Finally, this algorithm provides the means to carry out the reoptimized solution by executing step 2e in Table 10.2, in the real network. The risks involved in step 2e are limited, since only one demand is rerouted at a time, and the operation does not impact on the service if backup reoptimization is used.

10.4 The Complexity of Reoptimization

We have seen in Chapter 5 that the optimum routing of SBPP demands is a very difficult problem (NP-hard) [107]. In this section we discuss cases where the online routing or reoptimization algorithms as defined in Section 10.3 fail to find the optimum solution [54]. We provide here the proofs of such cases by counterexamples, as well as problem instances that can be used as comparison points to compare different optimization algorithms.

In the presentation of the proofs we use the following notation:

- P is an instance of an SBPP routing problem, it consists of a prescribed capacited network, point-to-point demands, and protection type for each demand.

- $Sol(P)$ represents the set of all possible solutions of P, which includes the routing of each demand, and the channels used (and shared) by each demand.

- $Opt(P) \in Sol(P)$ is the subset of optimum solutions of P, solutions that utilize the minimum number of channels over all possible solutions, i.e. $cost \ Opt(P) = \min_P cost \ Sol(P)$.

Assume that we use an online routing algorithm to solve this problem. An online routing algorithm is any algorithm that satisfies all three conditions: (1) demands are routed in sequence, (2) routed demands are immutable, and (3) the primary-backup pair of every new demand is selected so that the resulting total number of channels is minimized.

- S is an ordered sequence of all demands in P, and $Sol(P, S) \in Sol(P)$ is the corresponding solution if an online routing algorithm is used in conjunction with this sequence.

- In addition, for an ordered sequence S, let $Reopt(P, S) \in Sol(P)$ designate the reoptimized solution $Sol(P, S)$.

In this section, a reoptimized solution is the result of applying the reoptimization algorithm described in Table 10.2 on an existing solution, which is itself obtained using an online algorithm, possibly after trying all possible sequences.

Lemma 1: By definition of optimality, the cost of an optimum solution is the minimum over the costs of all possible solutions, including solutions found using an online routing algorithm with demands routed in any sequence ($cost \ Sol(P, S) \geq cost \ Opt(P) \ \forall S$).

10.4.1 No Prior Placement of Protection Channels or Primary Paths

In this section we prove three basic constructs on optimality of online routing and reoptimization in the case where no prior placement of protection channels or primary paths exists [54].

Assertion 1:

There are network instances for which no sequence exists for online routing that can achieve the optimum routing configuration.

$\exists P$ so that $cost \ Sol(P, S) > cost \ Opt(P) \ \forall S$

Proof:

With the help of Figure 10.2 we demonstrate the existence of at least one instance P for which the assertion is true. Part (i) of the figure illustrates P_0, a 12-node network, with two demands (a, b) and (c, d). We solve P_0 using an online routing algorithm and all possible sequences $S_1 = \{(a, b); (c, d)\}$ and $S_2 = \{(c, d); (a, b)\}$. Parts (ii) and (iii) of the figure depict two possible solutions. The dotted lines in the solution represent shareable protection channels. The solution in part (ii) could result from either sequence S_1 or S_2. The other example shown in part (iii) results from sequence S_1 only. There are other solutions not shown in this figure; we show, however, from the definition of the online routing algorithm that in this particular example, the cost is the same for all solutions of each

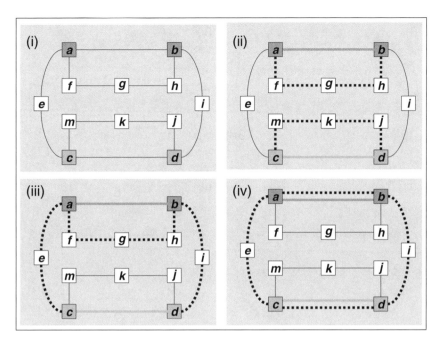

Figure 10.2: (i) Network with demands (a, b) and (c, d). (ii), (iii) Two sub-optimum solutions using best sequence of online routing. (iv) An optimum solution. (From [57], Figure 3. Reproduced by permission of © 2002 The NOC.)

given sequence S_i. The symmetry of the network guarantees the independence of the order of the sequence, and the first demand is always routed along the single hop primary and its corresponding four-hop backup path. Because the objective of the algorithm is to minimize the number of channels required for each new demand, and there is only one possible configuration for the first demand, the cost of routing the second demand is the same for all possible solutions resulting from this algorithm. Now we find that the minimum of cost $Sol(P_0, S_1)$ and cost $Sol(P_0, S_2)$ requires 2 primary channels, and 8 channels are reserved for protection, that is, a total of 10 channels. In comparison, the optimum solution $Opt(P_0)$ shown in part (iv) of the figure requires 2 primary channels and 6 channels are reserved for protection–therefore for this example $cost\ Sol(P_0, S) > cost\ Opt(P) \forall S.$ ∎

Using the same example as given in Figure 10.2, we can demonstrate a similar result for the reoptimization algorithm, in which the following assertion applies:

Assertion 2:

There are network instances for which no reoptimization exists that can achieve the optimum routing configuration.

$\exists P$ so that $cost\ Reopt(P, S) > cost\ Opt(P)\ \forall S$

Proof:

Removing any demand (a, b) or (d, c) from case (ii) or (iii) in Figure 10.2 and attempting to reroute it would only achieve the same result.

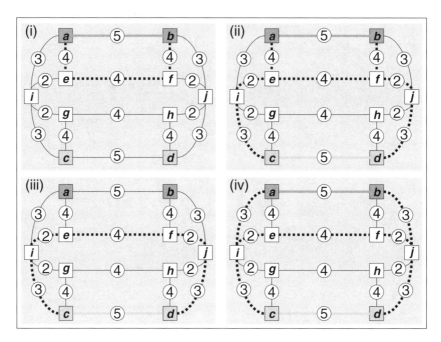

Figure 10.3: (i) Network with demand (a, b) routed and demand (c, d) to be routed. (ii) Min-cost routing of demand (c, d) assuming configuration (i). (iii) Configuration with demand (a, b) removed, and (iv) Configuration with demand (a, b) rerouted. (From [54], Figure 6. Reproduced by permission of © 2005 The Institute of Electrical and Electronics Engineers.)

In this particular case the reoptimization did not bring further improvement with respect to the online algorithm. This result is not to be generalized, however. And in fact we prove in the next assertion that reoptimization can achieve the optimum result, while there exists no sequence for which an online routing algorithm can. ■

Assertion 3:

There are network instances for which a reoptimization exists that can achieve the optimum routing configuration, while no sequence exists for which online routing can achieve the optimum routing configuration.

$$\exists P \text{ so that } cost\ Sol(P, S) > cost\ Reopt(P, S) = cost\ Opt(P)\ \forall S$$

Proof:

With the help of Figure 10.3 we demonstrate the existence of at least one instance P for which the assertion is true. Part (i) of the figure illustrates P_1, a network with two demands (a, b) and (c, d). Unlike the previous example where the cost was identical for every link, the links of this example may traverse a varying number of adjacent channels and intermediate degree-2 nodes. The circled values on the links represent the number of such channels (hops) and are used to indicate the effective cost of using the corresponding links. Part (i) of the figure shows the min-cost routing of demand (a, b). Part (ii) shows the subsequent min-cost routing of demand (c, d) following the routing given in part (i). Had the two demands been routed in the reverse order, an analogous solution would have been obtained with the routes of demands (a, b) and (c, d) reversed. Note that these are the solutions returned by the online routing algorithm, using any possible sequence (a, b), (c, d) or (c, d), (a, b). These solutions require 10 working channels and 22 protection channels. Applying the reoptimization algorithm to the solution of part (ii), we remove demand (a, b), as shown in part (iii), and reroute it using the min-cost algorithm, resulting in the solution shown in part (iv). The latest solution requires 10 working channels and 20 protection channels, an improvement of two channels compared to the best possible solution obtained by way of the online routing algorithm. By inspection we can show that this is the minimum cost achievable for this network. ■

10.4.2   Prior Placement of Protection Channels or Primary Paths

In this section, we prove three constructs on optimality of reoptimization and online routing with prior placement of protection channels or primary paths [54].

As an extension to Assertion 1, we can show that knowledge of the optimal placement of the protection channels is insufficient to determine the optimum solution using either the online routing or the reoptimization algorithm. This is the topic of the next two assertions. We begin these assertions with an extension of our terminology. For any instance P, let P^+ denote the instance with the protection channels optimally placed.

Assertion 4:

Even if the shared channels, part of the optimal routing configuration, are given, there are instances for which no online routing sequences exist that achieve the optimal solution.

$\exists P$ so that $cost\ Sol(P^+, S) > cost\ Opt(P)\ \forall S$

Proof:

Again, we demonstrate this assertion by way of an example. Using the network given in Figure 10.4, part (i) of the figure illustrates a 13-node network used to transport three demands: one demand between (c, d) and two demands between (a, b). We introduce demand (c, d) in order to limit the number of optimum solutions to one possible configuration, shown in part (ii) of the figure. Assuming next that the protection channels are allotted according to this optimum configuration as shown in part (iii), we determine by inspection that regardless of the sequence order, there are only three possible solutions achievable by the online routing algorithm. All three solutions require the same total number of channels. The solution that requires the least working capacity is shown in part (iv) of the figure. Comparing the optimum solution depicted in part (ii) with the solution in part (iv), we observe that the latter requires three more channels, which confirms the assertion. ■

Using the same example as given in Figure 10.4, we can demonstrate a similar result for the reoptimization algorithm, and the following assertion applies:

Assertion 5:

Even if the shared channels, part of the optimal routing configuration, are given, there are instances for which no reoptimization exists that achieves the optimal solution.

$\exists P$ so that $cost\ Reopt(P^+, S) > cost\ Opt(P)\ \forall S$

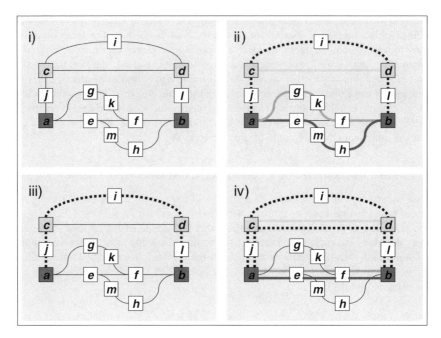

Figure 10.4: Shared backup path protection architecture. (i) Network with two demands (a, b) and one demand (c, d), (ii) An optimum solution, (iii) optimum set of channels reserved for protection, and (iv) suboptimum solution obtained from ii) using a greedy online routing algorithm. (From [54], Figure 7. Reproduced by permission of © 2005 The Institute of Electrical and Electronics Engineers.)

Proof:

Removing any demand (a, b) or (c, d) from case (iv) in Figure 10.4 and attempting to reroute it would only achieve the same result. ■

To conclude this section, we show that even if the primary paths are given, there are problem instances for which there is no sequence or reoptimization of the demand that achieves the optimum solution. In the proof of this assertion, for any instance P, let P^* denote the instance with the optimal primary paths.

Assertion 6:

Even if the primary paths, part of the optimal routing configuration, are given, there are instances for which no online routing sequence exists that achieves the optimal solution.

$$\exists P \text{ so that } cost\ Sol(P^*, S) > cost\ Opt(P)\ \forall S$$

Proof:

The proof is derived directly from the example of Figure 10.2. We observe that in this example the sequential algorithm, or reoptimization algorithm, always used the optimum primary paths, but failed to find the optimum backup paths. Thus, prescribing the primary paths in this example would result in the same suboptimum solution. ■

The proof is also valid for the reoptimization algorithm, and hence our final assertion:

Assertion 7:

Even if the primary paths, part of the optimal routing configuration, are given, there are instances for which no reoptimization exists that achieves the optimal solution.

$$\exists P \text{ so that } cost \; Reopt(P^*, S) > cost \; Opt(P) \; \forall S$$

Proof:

Using case (ii) or (iii) of Figure 10.2, we observe that removing either demand (a, b) or (c, d) and trying to reroute it would only achieve the same result. ∎

In conclusion, while it is sometimes impossible to achieve optimum routing with either online routing or reoptimization, there are cases where it can be achieved through reoptimization, while sequential routing only cannot achieve it (Assertion 3). The reverse is not true, since the first iteration of reoptimization is by definition a sequential algorithm, and any subsequent iteration is an improvement of the first iteration. We will show in the next section that the reoptimization algorithm achieves bandwidth savings in many circumstances.

10.5 Experiments

10.5.1 Calibration

In the following experiments, all the randomly generated graphs consist of rings traversed by chords connecting randomly selected pairs of nodes. Very often, but not always, it is possible to embed such a ring on a real network (the embedding requires finding a Hamiltonian circuit in the network), as demonstrated in Figure 10.5 with the ARPANET network. Each link of the random networks is assigned an arbitrary weight that represents the cost of using a channel in it. We then compare the algorithms according to their ability to minimize the total network cost, that is the number of required channels weighted by their respective link costs.

We first apply the complete reoptimization algorithms to small random generated networks, varying in size (i.e. number of nodes), with demands preliminarily routed using the online routing algorithm, and compare the solutions with results obtained by way of an ILP solver (see formulation in Section 5.6). The ILP solver is CPLEX 7.1[161] from ILOG. CPLEX exploits a branch-and-cut algorithm, in which it solves linear subproblems after setting a subset of formulation variables to integer values. In this process the ILP produces progress reports according to the solutions it finds: lower bounds if some of the solution's variables have fractional values, or upper bounds if the solution is feasible and all its variables have integer values. When the upper and lower bounds finally converge, the solution is known to be feasible and optimal. However, the convergence time of this process can be exponential to the size of the problem. We therefore put a limit of 10 hours maximum on each problem, and interrupted the process if no feasible solution could be found within that time frame, in which case not the optimum value, but a lower bound of it was used in the results of our experiments. Our observations, summarized in Table 10.3, indicate that for these networks, reoptimization allows capacity savings of 2–6%, and is within 1% of the optimal solution, or 2% of a known lower bound if the optimal solution cannot be determined.

10.5.2 Real Networks

Next, we apply the algorithm to reoptimize the routes of four different networks, Net-A, Net-B, Net-C and Net-D. Net-A is a network built by Dynegy [66]. It consists of 48 nodes, 75 links, and a number

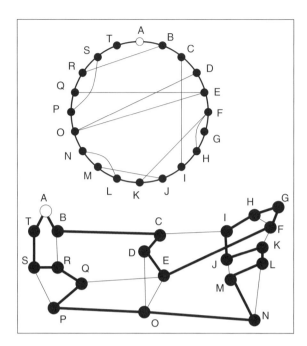

Figure 10.5: Chordal ring (top) embedded on ARPANET (bottom). (From [54], Figure 8. Reproduced by permission of © 2005 The Institute of Electrical and Electronics Engineers.)

$Q \approx 100$ (order of hundred) of SBPP demands with their routes provided by the operator. Net-A consists of three periods (Net-A.1, Net-A.2 and Net-A.3), measured over a 14-month interval, that captured the actual growth of the network. This scenario has a limited number of spare channels, and offers very little room for rearranging the paths. Net-B consists of 25 nodes, 30 links and 290 demands, Net-C is a 45-node network with 75 links and 570 demands, and Net-D is a 60-node network with 90 links and 195 demands. The characteristics of the networks are summarised in Table 10.4.

The demands of Net-B, Net-C and Net-D are provided unrouted with source and destination information only. Henceforth, we created an initial routing configuration for these three scenarios by routing their demands sequentially following an arbitrary order, using the Compute_Pair_of_Paths online routing procedure described in Section 10.2. We added new channels as needed during that process, assuming that the network had infinite capacities. The demands of each scenario are then reoptimized, exercising both the complete and the backup only reoptimizations, using the Reopti-mize_Demands procedure of Section 10.3.

Using networks Net-B, Net-C and Net-D we perform a series of experiments assuming two different scenarios. In the first scenario, referred to as the Static Network Infrastructure, we assume that no capacity is added, or freed by demand churn, from the moment the demands are routed the first time, and the moment the reoptimization is executed. In the second scenario, referred to as the Growing Network Infrastructure, we assume that as the network grows and demands are routed, new links or new channels on existing links are added to the network, creating opportunities for improvement of existing demands.

The case of network Net-A was treated as a growing network infrastructure only, since this is the context in which this network has been built.

Table 10.3: Comparison of complete reoptimization with ILP-based solution. (From [54], Table 1. Reproduced by permission of © 2005 The Institute of Electrical and Electronics Engineers.)

| | Number of | | Online routing algorithm | | |
Nodes	Links	Demands	Working channels	Backup channels	Total cost
10	13	90	404	392	5624
12	15	99	636	568	9568
15	19	95	500	358	6786

| | Number of | | Reoptimized (complete) | | |
Nodes	Links	Demands	Working channels	Backup channels	Total cost
10	13	90	396	376	5428
12	15	99	552	570	8918
15	19	95	514	326	6626

| | number of | | ILP Solver (CPLEX) | | |
Nodes	Links	Demands	Working channels	Backup channels	Total cost
10	13	90	396	376	5428
12	15	99	552	566	8894
15	19	95	(1)	(1)	3476

Table 10.4: Characteristics of networks used in experiments. (From [54], Table 4. Reproduced by permission of © 2005 The Institute of Electrical and Electronics Engineers.)

Network	Number of nodes	Number of links	Number of demands
Net-A.1	48	75	$Q \approx 100$
Net-A.2	48	77	$1.06 \times Q$
Net-A.3	48	79	$1.42 \times Q$
Net-B	25	30	290
Net-C	45	75	570
Net-D	60	90	195

10.5.3 Static Network Infrastructure

Tables 10.5 and 10.6 summarize the results for the backup and the complete reoptimization respectively when the infrastructure of the network remains static throughout the experimentation. The tables show the quantities measured before and after reoptimization. For each scenario, the same network and routed demands are used for backup and for complete reoptimizations. The number of ports in Table 10.5 consists of ports used for the protection channels only, since the working channels remain the same. The number of ports in Table 10.6 consists of all the ports in the network, used for primary and

Table 10.5: Backup only reoptimization–static network infrastructure. (From [54], Table 2. Reproduced by permission of © 2005 The Institute of Electrical and Electronics Engineers.)

Scenario	Backup port count				Average backup hops		Max. backup hops	
name	Before	After	% saved	% of total ports saved	Before	After	Before	After
Net-B	2520	2452	3	1	5.83	5.76	11	10
Net-C	2340	2242	4	2.1	4.61	4.53	15	13
Net-D	504	470	7	3	4.37	4.10	11	9

Table 10.6: Complete reoptimization–static network infrastructure. (From [54], Table 3. Reproduced by permission of © 2005 The Institute of Electrical and Electronics Engineers.)

Scenario	Total (working and backup) ports			Backup port count			Average backup hops		Max. backup hops	
name	Before	After	% saved	Before	After	% saved	Before	After	Before	After
Net-B	5088	4994	2	2520	2502	0.71	5.83	5.72	11	10
Net-C	4640	4450	4	2340	2174	5.48	4.61	5.20	15	16
Net-D	1112	1036	7	504	470	6.75	4.37	4.25	11	13

protection channels. We observe that backup only reoptimization saves up to 3% of the total number of ports, and complete reoptimization up to 7%. The complete reoptimization offers the most cost-efficient alternative, but most of the improvement is realizable using the backup only reoptimization algorithm, without service interruption.

We also observe that the protection path latency tends to be slightly longer in complete reoptimization than in backup only reoptimization. Although counterintuitive at first, this is actually an expected outcome in the case of SBPP networks, because the complete reoptimization algorithm explores a wider solution space in which backup paths are slightly longer on the average than in backup reoptimization. This effect can be mitigated if necessary by increasing the value of ϵ, as indicated earlier in the description of the cost model.

10.5.4   Growing Network Infrastructure

The next set of experiments covers the case of the growing network infrastructure. Here demands are routed online over time, while new links and capacity on existing links are added simultaneously to keep up with the demand growth, creating more realistic network dynamics than the previous exercise. For the case of Net-B, Net-C and Net-D, we first route the demands after removing a link selected empirically by inspection. In particular, we favor a link that exhibits an apparent impact on the network connectivity, but is not essential to protect all the demands. For instance in the example of Figure 10.5, a good candidate link for removal would be $\{E, F\}$ or $\{O, N\}$. The link is then reinserted to simulate a capacity upgrade, and the demands routed in the first step are reoptimized in the upgraded network. Actual network Net-A was built upon the gradual addition of such strategic links. For instance, Table 10.4 indicates that the size of Net-A increased from 75 to 79 links over the study period. Other affecting factors are the policy used to add new channels with respect to demand growth, and the pattern of channel unavailability caused by maintenance or failure conditions. Tables 10.7 and 10.8 summarize the results for the backup and the complete reoptimization respectively when the infrastructure of the network evolves over time. As before, the tables show the quantities

Table 10.7: Backup reoptimization–growing network infrastructure. (From [54], Table 5. Reproduced by permission of © 2005 The Institute of Electrical and Electronics Engineers.)

Scenario name	Backup port count				Average backup hops		Max. backup hops	
	Before	After	% saved	% of total ports saved	Before	After	Before	After
Net-A.1	224	208	7	4	7.1	5.24	20	11
Net-A.2	310	214	31	19	5.9	4.88	15	10
Net-A.3	332	242	27	15	4.8	3.57	13	9
Net-B	2986	2868	4	2	6.33	6.02	10	10
Net-C	2576	2294	11	5.5	5.07	4.73	15	15
Net-D	550	520	5	2	5.10	4.13	16	13

Table 10.8: Complete reoptimization–growing network infrastructure. (From [54], Table 6. Reproduced by permission of © 2005 The Institute of Electrical and Electronics Engineers.)

Scenario name	Total (working and backup) ports			Average backup hops		max. backup hops	
	Before	After	% saved	Before	After	Before	After
Net-A.1	400	382	5	7.1	5.43	20	10
Net-A.2	494	372	25	5.9	4.82	15	10
Net-A.3	612	486	21	4.8	4.05	13	9
Net-B	5306	4970	6	6.33	6.02	10	11
Net-C	5134	4564	11	5.07	4.80	15	14
Net-D	1326	1030	22	5.10	4.61	16	13

measured before and after reoptimization. Note that the savings for Net-A are substantial. Unlike the static scenarios where channel availability is not an issue and assumed to be unlimited during the initial routing, the demands of this network have been routed while new channels were being added later, or while existing channels or links may have been unavailable for maintenance reasons, thus creating opportunities for optimization. This is the most realistic mode of operation, and the most likely to occur. Worth noticing for this scenario is the reduction in protection path latency measured as the average number of channels traversed by the protection paths, which decreases from 7.1 to 5.24 hops for the backup reoptimization of Net-A.

Note that depending on the scenario, the difference in terms of performance between backup only and complete reoptimization is more or less pronounced. This can be due to a combination of factors, such as the demand set or the network topology. Most importantly, it is how network capacity and demand growth are achieved. In the case of Net-A, the two occur simultaneously and this case should be considered separately. In the case of Net-B, Net-C and Net-D, it depends on the link that is added to simulate growth. We can illustrate this with a network constituting two regions R1 and R2 only connected by two fibers F1 and F2. Any pair of demands between R1 and R2 cannot share backup capacity in F1 or F2, because in this part of the network either their working paths are not disjoint, or their backups are disjoint. If we add a third fiber F3 between R1 and R2, and reoptimize the demands between the two regions, the new capacity can be used to enable sharing across the backup paths. However, in order to get the most benefit, it may be necessary to reoptimize the working paths as well. For instance, if the working paths are all routed on F1 by the initial online routing, then the addition of fiber F3 followed by a backup reoptimization does not enable capacity sharing between the backups, because all the working paths have a common point of failure on fiber F1. On the other hand, a complete reoptimization that distributes the working paths on the three fibers would allow sharing.

10.5.5   Network Dynamics

We had the opportunity to observe the effect of reoptimization on Net-A at different intervals over a 14-month period. Net-A is the Dynegy network implemented using Tellium's Aurora^{TM} optical cross-connect and SBPP architecture [65, 66]. Our measurements shown in Figure 10.6 demonstrate that the network imitates the behavior illustrated in Figure 10.1. To conclude the study with a real-life experiment, two actual backup reoptimizations of the network were performed over this 14-month period on the Dynegy network, each time saving the operator 27–31% of the transport cost by rerouting a subset of the backups. Each time the actual reoptimization took place overnight, changing the lightpaths from protected state to unprotected through a provisioning command, and then from unprotected back to protected through another provisioning command that would compute a new backup path. The sequence of lightpaths was computed using the algorithm shown in Table 10.2. The freed capacity was then immediately available to carry new demands.

10.6   Conclusion

In this chapter we have presented a reoptimization algorithm to rearrange shared backup path-protected lightpaths. The proposed algorithm allows for two types of reoptimization. A complete reoptimization algorithm that reroutes both primary and backup paths, and a backup reoptimization algorithm that reroutes the backup paths only. Rerouting backup paths only is a suboptimal but attractive alternative that avoids any service interruption. Our simulations indicate that even on properly planned networks, the complete reoptimization achieves 3–5% savings in the cost of the transport, and most of the improvement can be achieved by way of the backup reoptimization alone. In real-life scenarios,

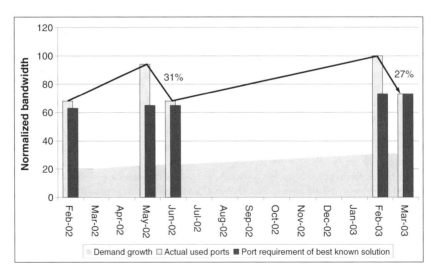

Figure 10.6: Network growth over a period of 14 months, and the effect of backup reoptimization on backup bandwidth utilization. (From [54], Figure 9. Reproduced by permission of © 2005 The Institute of Electrical and Electronics Engineers.)

where the demand and the capacity of the network grow over time in a more unpredictable manner, the cost savings are of the order of 30% of the cost of the network. In this chapter we also proved that although none of the proposed algorithms is guaranteed to find the minimum cost solution, there exist cases where the reoptimization algorithm can achieve the optimum result, while for the same cases there are no demand sequences for which a sequential routing algorithm can.

Chapter 11

Dimensioning of Path-Protected Mesh Networks

11.1 Introduction

While designing or dimensioning an optical transport network it can be valuable to quickly obtain a gross estimate of its capacity and resulting cost, and a sense of its performance. Often the information provided and/or the time available to complete the task at hand may be insufficient to proceed with a full-scale study of the network, using a formal optimization method such as ILP-based optimization (for capacity planning) and running a simulation (for performance analysis). The task is further hindered by increasingly complex failure-recovery architectures which affect both capacity (and therefore cost), and performance. For instance, with Shared Backup Path Protection, additional capacity is reserved to secure for every demand an alternate route that serves as backup in case of failure occurrence along the primary route [93, 99, 107, 179, 201]. Since not all demands will be affected by a single failure, the reserved capacity can be shared among multiple demands. The amount of sharing that can be achieved directly impacts on the total network cost. The time to seize the shared capacity and reestablish service after a failure is a critical network performance characteristic. Both metrics can be complex and time-intensive to compute. The objective of this chapter is to present some recently proposed analytical and semiempirical approaches by Labourdette *et al.* [188, 190] (Sections 11.3 and 11.5) and Korotky and Bhardwaj [41, 42, 43, 183, 184, 185] (Section 11.6) for developing models and deriving sets of formulas to quickly estimate these metrics.

Formulations are based on classes of networks and demand models that tend to be idealized (e.g. random networks, regular networks, uniform demands) and usually allow for analytical modeling not readily possible with real networks and arbitrary demand sets. In particular, these idealized networks may not have the practical constraints that real networks have (such as proximity constraints whereby a switch can only be connected to a subset of neighboring switches within a certain radius). The consequence for those networks is a more uniform and compact inter-nodal connectivity. Nevertheless, the hope is that the proposed formulations can be used to assess network capacity (and cost), and network performance, at least in a relative sense, and pinpoint design deficiencies.

Recently, work by Forst and Grover [123] presented an extensive analysis of the analytical estimates by Korotky and Bhardwaj [183] and Labourdette *et al.* [190]. In their work discussed in Section 11.7, the authors of [123] conducted a further series of tests of the equations to assess their

Path Routing in Mesh Optical Networks   Eric Bouillet, Georgios Ellinas,
Jean-François Labourdette, Ramu Ramamurthy   © 2007 John Wiley & Sons, Ltd

accuracy over a more general and extensive set of network topologies and demands than the ones used to develop the models. This study identifies some of the limitations of a pure analytical model compared to a formal optimization method.

11.2   Network and Traffic Modeling

In what follows, we represent a WDM network as a graph. Vertices (or nodes) represent the optical switches in the network, and edges represent (bidirectional) optical links. We use n and m to denote respectively the number of vertices and edges. We call degree of a vertex the number of edges terminating at this vertex. The minimum, maximum and average vertex degrees of a graph are respectively denoted $\delta_{\min}$, $\delta_{\max}$ and δ. It is easily shown that $\delta = 2m/n$. A path $p(a, z)$ is a succession of distinct edges in tandem traversing distinct vertices starting with vertex a and ending with vertex z. The hop length of a path is the number of successive edges that constitute the path. There may be many paths between any vertex pair (a, z). Among all these paths we call min-hop path(s), the subset of path(s) with minimum hop length, and call $h(a, z)$ this minimum hop length. The diameter D of a graph is the longest of all minimum hop lengths calculated among all the vertex pairs, i.e. $D = \max_{(a,z)} h(a, z)$. In the remainder, we assume that all SRGs are of the default type (denoted as type (a) in Chapter 3) with one SRG per link and one link per SRG. We also assume no parallel links.

In addition, we represent the traffic between the different network nodes as a matrix $T = \{t_{ij}\}$ of demand t_{ij} between nodes i and j in the graph representation. We let h_{ij} represent the minimum hop length between nodes i and j. The average minimum hop length can be computed exactly as the demand-weighted average of the minimum hop lengths among all node pairs, $h = \sum_{i,j} t_{ij} h_{ij} / \sum_{i,j} t_{ij}$. In the remainder of this chapter the traffic demand is generally assumed to be uniform unless otherwise noted. This means that there is equal demand between all node-pairs in the network, $t_{ij} = t, \forall i, j$. It follows that under uniform demand, $h = \sum_{i,j} h_{ij} / (n(n-1)/2)$, assuming an uncapacitated network. Under capacity constraints, the model becomes more complicated as multiple demands between two nodes could be routed differently when the link or node capacity constraints are reached, possibly yielding different paths and hop lengths between node-pairs. Different demand models would change the set of results. If all the demands were between adjacent (connected) nodes, it would follow that $h = 1$.

Finally, a cost framework could be developed which associates specific costs with different components of the network, typically node and link costs. As costs can change rapidly with the introduction of new technologies and the maturing of established technologies, we will work in the remainder of this chapter with bandwidth metrics. The bandwidth metrics of interest addressed later in this chapter are the node bandwidth, i.e. number of ports, and the link bandwidth, i.e. number of WDM channels. Doing so allows us to develop a general framework where costing can be later applied based on the state of the telecommunications equipment industry and the introduction of new technologies. A number of papers have studied the impact of equipment and node costs, both of the switching, transport and IP routing components, on the optimization of networks, for example [183] for overall network costs, [185, 189, 229] for two-tier networking optimization, and [186, 276] for IP-over-DWDM network design.

11.3   Mesh Network Characteristics

In this section, we will analyze and develop approximate formulas for the length of primary and backup paths for Dedicated Backup Path Protected (DBPP) and Shared Backup Path Protected (SBPP) connections. We will also model the ratio of protection to working capacity. In particular, we will present a mathematical approach to estimate the amount of sharing in SBPP networks. This will allow us to estimate the amount of protection capacity required and the resulting ratio of protection to working capacity in a mesh network. Throughout the section, we assume that the traffic demand is uniform.

11.3.1 Path Length Analysis

The well-known Moore bound [40] gives the maximum number of nodes in a graph of diameter D and maximum degree δ_{max}:

$$n \le 1 + \delta_{max} \sum_{i=1}^{D} (\delta_{max} - 1)^{i-1} \tag{11.1}$$

$$n \le 1 + \delta_{max} \frac{(\delta_{max} - 1)^D - 1}{\delta_{max} - 2}, \delta_{max} > 2 \tag{11.2}$$

As illustrated in Figure 11.1, the Moore bound results from the construction of a tree whose root is the parent of δ_{max} vertices and each subsequent vertex is itself the parent of $\delta_{max} - 1$ vertices. The underlying idea is to pack as many vertices in D generations (hops) as is possible with respect to δ_{max}. The bound implies the existence of one such tree growing from every vertex and embedded in the graph, and is thus difficult to attain. It is nevertheless achievable for rings with odd number of vertices and for fully connected graphs. Reciprocally, given the number of nodes n, and maximum degree δ_{max}, the lower bound D_{min} on the graph's diameter is easily obtained from Equation 11.2:

$$D_{min} \ge \frac{\ln\left[(n-1)\frac{\delta_{max}-2}{\delta_{max}} + 1\right]}{\ln(\delta_{max} - 1)}, \delta_{max} > 2 \tag{11.3}$$

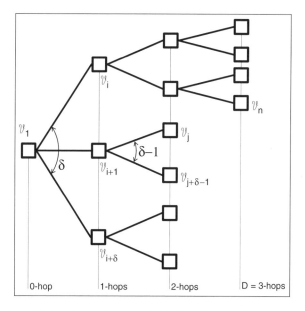

Figure 11.1: Moore tree. First vertex v_1 is connected to δ vertices one hop away from v_1. Subsequent vertices are connected to $\delta - 1$ vertices one hop farther from v_1. (From [190], Figure 16. Reproduced by permission of © 2005 The Institute of Electrical and Electronics Engineers.)

Equations 11.1 and 11.3 can be combined to determine a lower bound on the average hop length h:

$$(n-1)h \geq \delta_{\max} \sum_{i=1}^{D_{\min}-1} i(\delta_{\max}-1)^{i-1} + D_{\min}\left(n-1-\delta_{\max}\sum_{j=1}^{D_{\min}}(\delta_{\max}-1)^{j-1}\right) \qquad (11.4)$$

Equation 11.4 is a rather conservative lower bound of the average path length. Instead of this expression we will use Equation 11.5 below.

$$h \approx \frac{\ln\left[(n-1)\frac{(\delta-2)}{\delta}+1\right]}{\ln(\delta-1)}, \delta > 2 \qquad (11.5)$$

Equation 11.5 is exact for complete mesh networks ($h \to 1$ when $\delta \to n-1$), it behaves as expected for infinite size networks ($h \sim \ln n / \ln(\delta-1)$ when $n \to \infty$), and our experiments indicate that it gives a fair approximation of the average path length for the idealized networks as discussed earlier. Note that this is an approximation as one may want to take a longer working path than the shortest path either (a) to be able to find a diverse backup path in the case of DBPP or SBPP lightpaths, or (b) to maximize sharing in the case of SBPP lightpaths. However, it is our experience and it is reasonable to expect that the length of the shortest path gives a very good approximation for the length of the working path for both DBPP and SBPP lightpaths.

The computation of the average hop length of the backup path $a - z$, in the context of Dedicated Backup Path Protection, is easily derived from a trivial transformation of the graph as shown in the example of Figure 11.2. The transformation consists of (1) removing edges on the primary path, and (2) because we assume no parallel edges, selecting b, a neighbor of a, removing node a, and removing edge (a, b). The purpose of the transformation is to determine the average degree δ' of the new graph (the number of nodes is $n - 1$). We then compute the average hop length h_{bz} of path (b, z) on the modified graph using Equation 11.5. The average hop length of backup path (a, z) is $h_{az} = 1 + h_{bz}$.

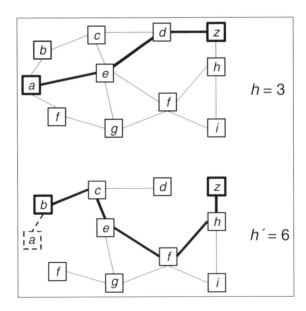

Figure 11.2: Graph transformation for the computation of the backup path length. (From [190], Figure 5. Reproduced by permission of © 2005 The Institute of Electrical and Electronics Engineers.)

Using the transformation explained above, the new average degree is

$$\delta' = \frac{2[m - (\delta + h) + 1]}{n - 1} \tag{11.6}$$

where h is the average hop length of the primary path as computed in Equation 11.5. Plugging back Equation 11.6 into Equation 11.5, the average hop length h' of the backup path can be approximated as

$$h' \approx \frac{\ln\left[(n-2)\frac{(\delta'-2)}{\delta-1} + 1\right]}{\ln(\delta' - 1)}, \delta > 2 \tag{11.7}$$

Note that Equation 11.7 contains both δ and δ'. This is because $\delta - 1$ in Equation 11.7 stands for the degree of the vertex origin of the path. Its degree is the average degree of the unmodified graph (used to compute the primary) minus one to account for the primary, hence $\delta - 1$.

Figure 11.2 shows an example of the transformation procedure. The backup path cannot traverse edges already used for its primary (a, z), and so these edges can be removed. Furthermore, the backup is at least as long as the primary and we assume no parallel edges, hence the backup is at least two hops long. This is represented by automatically adding one hop to the length of backup (a, z) and starting from any neighbor b of a, other than z.

Figures 11.3 and 11.4 plot the approximations for h and h' against experimental path lengths computed in randomly generated networks [190] with average node degrees 3 and 5 respectively, which encompass the representative range of real telecommunications networks. As seen from the plots, there is a very good match between the experiments and the approximation formulas for h and h'. Experiments on similar networks with varying degrees exhibit the same behavior.

In the case of Shared Backup Path Protection we introduce ϵ, the cost of a shareable channel reserved for protection [55, 97, 241]. The cost is actually the ratio of the cost of a shareable channel to the cost of the same channel if it were not shareable (see Chapter 7 and [55] for details). The ratio ϵ ranges from 0 to 1. Equation 11.8 is an empirical model for backup hop length h'' as a function of ϵ and a parameter h_0 for SBPP. In this model h is the length of the primary path (assumed to be the

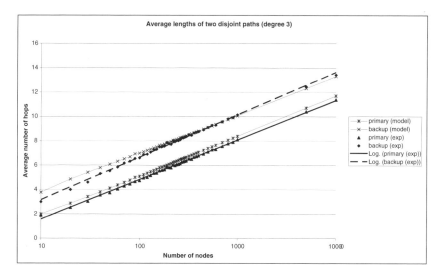

Figure 11.3: Approximated and measured path lengths (networks of degree 3). (From [190], Figure 6. Reproduced by permission of © 2005 The Institute of Electrical and Electronics Engineers.)

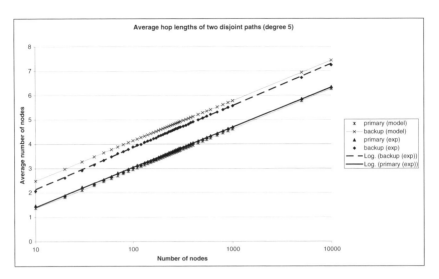

Figure 11.4: Approximated and measured path lengths (networks of degree 5).

same for both DBPP and SBPP lightpaths) and is computed using Equation 11.5, and h' is the length of the backup path for DBPP and is computed using Equation 11.7.

$$h'' = h' + (1 - \epsilon)h_0 \qquad (11.8)$$

h_0 is a measure of the difference between h' and h'' (induced by sharing) and is determined experimentally (e.g. in the range of 0 to a few hops) or can sometimes be derived from the topology. For example, $h_0 = 0$ for a ring topology, or more generally for any topology where there is only one diverse path from the primary path. Note that in the case of a ring, $h'' = h' = m - h$.

If $\epsilon = 0$ the backup path is routed with sharing in mind, resulting in perhaps longer routes, but requiring less new channels by reusing existing ones already reserved for protection. For $\epsilon = 1$, the routing procedure seeks to minimize the hop length only. The primary and backup hop lengths are therefore identical to the DBPP case. Sharing is ignored during path computation, but is still performed once the path has been computed.

11.3.2 Protection-to-Working Capacity Ratio Analysis

A recurrent question in the case of SBPP and DBPP techniques concerns their respective overload measured in terms of capacity reserved for protection. This figure of merit is often expressed as the ratio of protection to working capacity, with lower ratio signifying more capacity-efficient protection. We define this ratio as follows:

$$\mathbf{R} = \frac{total\ number\ of\ protection\ channels}{total\ number\ of\ working\ channels} \qquad (11.9)$$

If the answer to this question is trivial in the DBPP case with the leverage of Equations 11.5 and 11.7, it requires more thought in the SBPP case.

In the DBPP case, the ratio $\mathbf{R_d}$ of protection to working capacity is also the ratio of the average protection path length to the average working path length, and is independent of the number of lightpaths in the network.[1]

$$\mathbf{R_d} \simeq \frac{h'}{h} \geq 1 \qquad (11.10)$$

---

[1]Assuming an uncapacitated network.

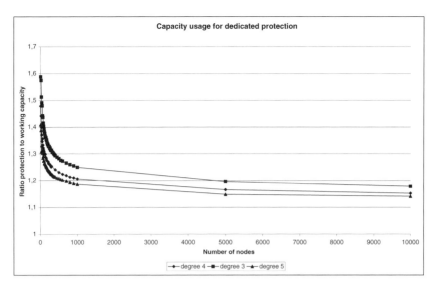

Figure 11.5: Ratio of protection to working capacity for DBPP.

Figure 11.5 shows how this ratio scales with large networks for node degrees ranging from 3 to 5. As expected, as the size of the network increases or the average node degree increases, the ratio becomes closer to 1 because there is a dedicated backup path (diverse from the primary or shortest path) whose length is only slightly longer than the shortest path.

In the case of SBPP, we cannot express $\mathbf{R_s}$ as the ratio of average protection path length to average working path length because some of the channels on a protection path can be shared between several lightpaths. We thus introduce a new parameter F, which represents the fill factor of a protection channel, that is the number of lightpaths whose protection paths are using that channel. Then, the ratio of protection to working capacity $\mathbf{R_s}$ can be expressed as

$$\mathbf{R_s} \simeq \frac{h''}{h}\frac{1}{F} = \mathbf{R_d}\frac{1}{F} + (1-\epsilon)\frac{h_0}{h}\frac{1}{F} \geq \frac{\mathbf{R_d}}{F} \qquad (11.11)$$

Note that $\mathbf{R_s}$, contrary to the DBPP case, is not independent of the number of lightpaths, as more lightpaths will provide for better sharing, thus increasing F, and therefore reducing $\mathbf{R}$.[2] However, F, and $\mathbf{R_s}$, should become independent of the number of lightpaths when that number becomes large enough.

11.3.3 Sharing Analysis

The sharing analysis consists of determining the relationship between F and the number of lightpaths, or demands, in a network. The analysis first determines the number of lightpaths whose backup path traverses an arbitrary link l, and then the largest number of corresponding primary paths that traverse any given link. That number is the number of backup channels required on the arbitrary link l, and F is simply the ratio of lightpaths whose backup path traverses l divided by the number of backup channels required.

In [190], the authors derive the average number of shared backup channels required on a link, as a function of the number of lightpaths L in the network, to guarantee recovery from single link failure.

---

[2]Again assuming uncapacitated networks.

The protection architecture used is that of pooling backup channels across all failures, that is, not preassigning channels to particular backup paths. The authors consider a network with n nodes and m links. The average node degree is $\delta = 2m/n$. The average length of the primary path of a lightpath is h, given by Equation 11.5. The average length of the backup path of a lightpath is h', given by Equation 11.7. The authors determine the maximum number of times primary paths of lightpaths whose backup paths share a given link l traverse a common link. Under pooling of shared backup channels, this is the number of backup channels needed on link l to insure that all lightpaths that would be subject to the failure of a common link can be recovered. The sharing factor is then simply the ratio of lightpaths whose backup paths traverse link l divided by the number of backup channels required.

Results from [190] comparing the value of $\mathbf{R_s}$ to this approximation are given here. Figures 11.6 and 11.7 compare the approximation of the sharing ratio against experimental sharing ratios computed in random chordal ring networks of 50 nodes and 75 links, and 150 nodes and 300 links respectively.

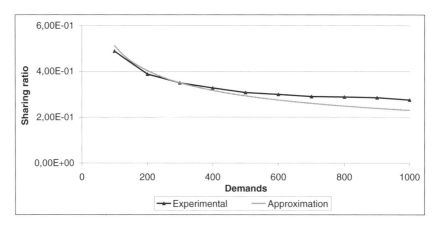

Figure 11.6: Sharing ratio, experimental vs approximation results as demand increases on a 50-node, 75-link chordal ring network (degree = 3). (From [190], Figure 8. Reproduced by permission of © 2005 The Institute of Electrical and Electronics Engineers.)

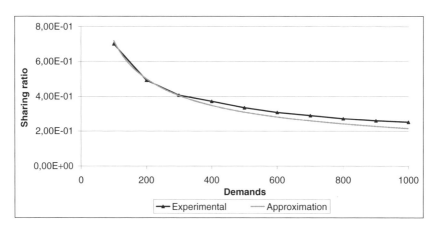

Figure 11.7: Sharing ratio, experimental vs approximation results as demand increases on a 50-node, 300-link chordal ring network (degree = 4). (From [190], Figure 9. Reproduced by permission of © 2005 The Institute of Electrical and Electronics Engineers.)

11.4 Asymptotic Behavior of the Protection-to-Working Capacity Ratio

In this section, we attempt to answer the following question: what happens to the protection-to-working capacity ratio $\mathbf{R_s}$ in the SBPP case for a topology with n nodes, average node degree δ, and a full mesh demand, as the number of nodes n increases to infinity?

11.4.1 Examples

First, we show how the capacity ratio $\mathbf{R_s}$ behaves for some simple network examples, namely a full mesh topology, and a bangle, and prove a nonconvergence to zero result when the degree remains fixed while the network size increases.

11.4.1.1 Full Mesh Topology ($\delta = n - 1$) (see Figure 11.8)

Assume $n - 2$ demands between each node-pair routed on the one-hop edge. The working capacity on each edge is $n - 2$. The protection capacity on each edge is 1. $\mathbf{R_s} = 1/(n - 2) \to 0$ as $n \to \infty$. For a full mesh topology, link protection and path protection are identical, and it is well known (see for example [287]) that for link protection, $\mathbf{R_s} \geq 1/(\delta - 1)$.

11.4.1.2 Bangle ($\delta = 3$) (see Figure 11.9)

There are two kinds of edges, ring edges (those on the rings), and tie edges, those edges that tie the rings. Let q be the number of tie edges, $2q$ the number of ring edges, so that $n = 2q$. Let W_1, P_1 be the working and protection capacity on ring edges, and W_2, P_2 be the working and protection capacity

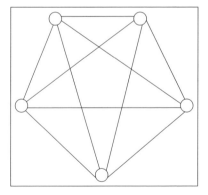

Figure 11.8: Full mesh topology ($\delta = n - 1$).

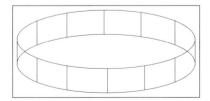

Figure 11.9: Bangle ($\delta = 3$).

on tie edges. Then P_1 and W_1 satisfy $P_1 \geq W_1/3$ by considering a cut of four ring edges. Also, P_2 and W_2 satisfy $P_2 \geq W_2/q$ by considering a cut of q tie edges. Finally, W_1 and W_2 satisfy $W_2/W_1 \leq 4/q$ since tie edges are used only once for a demand that goes from one ring to the other ring.

$$\mathbf{R_s} = \frac{P_1 + P_2}{W_1 + W_2} \geq \frac{W_1/3 + W_2/q}{W_1 + W_2} \geq \frac{\frac{1}{3} + 1/q}{1 + 4/q} \geq \frac{1}{6} + \frac{1}{2q} \geq \frac{1}{6} \qquad (11.12)$$

Therefore, for a bangle $\mathbf{R_s} \geq 1/6$ as $n \to \infty$.
We can extend the results for a bangle using the following theorem:

Lemma 1 *If δ is fixed independent of n, $\mathbf{R_s} \to \epsilon > 0$ as $n \to \infty$ for a class of topologies and demand sets.*

Proof. There is a cut in the topology such that the number of edges in the cut is limited to be $k\delta$ (some constant k). If the working capacity on the cut edge is W, then the protection capacity on each cut edge must be at least $W/k\delta$, and $\mathbf{R} \geq 1/k\delta$. ∎

In the next section, we generalize these results and consider the case where the average node degree increases with the size of the network according to certain rules.

11.4.2 General Results

Consider an optical network using SBPP. We make the following assumption about the optical network in an attempt to classify the topologies: there exists at least one Hamiltonian circuit that traverses all vertices in the graph exactly once. The example of Figure 11.10 illustrates such a graph and Hamiltonian circuit. It is easy to show that the Hamiltonian circuit is a ring with $m' = n$ edges, and a (nonsufficient) condition of existence is that all vertices have degree 2 or greater. We further assume no parallel edges. Note that the assumption of the existence of a Hamiltonian circuit has been used in [48] to derive properties of optimal survivable paths in mesh networks. We also assume unlimited protection capacity in order to allow unrestricted configuration of the protection overlay architecture and achieve optimality. The working capacity is bounded and we denote by $Q_{\min}$ and $Q_{\max}$ respectively the minimum and the maximum working capacity of any edge in the network. As usual, we denote by n and m respectively the number of vertices and the number of edges in the network. We denote $\delta_i = m_i$ the degree, i.e. number m_i of incident edges, of vertex v_i and $\delta = (1/n) \sum_i \delta_i = 2m/n$ the average degree of the network. We will make additional assumptions about the demand set, such as the minimum and the maximum number of hops of their corresponding

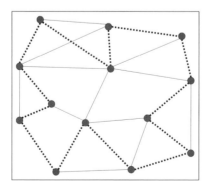

Figure 11.10: Graph $G(V, E)$ and example of Hamiltonian circuit (dotted lines).

working path set, or optimum capacity utilizations; these assumptions are detailed in the theorems that follow.

Theorem 1 *If we assume Q_{max}/Q_{min} constant, and the diameter of a telecommunications network to be of the order of $O(\log_\delta n)$ hops, then noting that the average node degree is $\delta = 2m/n$ we have for all graphs G, for which a Hamiltonian circuit H_c exists:*

1. *The upper bound of the protection-to-working capacity ratio for these graphs converges to 0 when n increases to infinity, if the average vertex degree δ of the graph strictly increases with n.*

2. *Otherwise, the protection-to-working capacity ratio for these graphs is a value in the range*

$$\left[\frac{Q_{min}}{Q_{max}} \frac{1}{\delta}, \frac{Q_{max}}{Q_{min}} \frac{O(\log n)}{\delta} \right]$$

   and it thus does not converge to 0.

We now derive a theorem for each of a series of independent assumptions.

Theorem 2 *Assume that all edges have working capacity 1, and infinite capacity for protection. Then there exists a demand set D so that the throughput is maximized, and the optimal solution uses all the working capacity with the ratio of protection-to-working capacity $\mathbf{R_s} = n/m = 2/\delta$.*

Proof. Consider a demand set D isomorphic to E, i.e. for each edge of E there is in D a corresponding one-unit demand between the two vertices incident to that edge and reciprocally (i.e. there is a one-to-one correspondence between the edges in E and the demands in D). Assume that we route each demand on a one-hop path along the edge itself. Our primary assumption is that there exists a Hamiltonian circuit traversing all the vertices of $G(V, E)$. Let $H_c(V, E)$ denote such a Hamiltonian circuit. For any edge (u, v) in E, after removing the edge there is at least one but no more than two paths $u - v$ in $H_c(V, E)$–one path if (u, v) is traversed by $H_c(V, E)$, and two paths if (u, v) is not traversed by $H_c(V, E)$. Suppose that we reserve one protection channel on every edge of $H_c(V, E)$, then these reserved channels are sufficient to protect the set D. Indeed, the primaries of the demands are mutually disjoint so that all can share a common set of protection channels, and there is a path of protection channels along $H_c(V, E)$ between any pair of vertices. Therefore, for this demand D, there is a solution that requires m working channels, and n protection channels. We now show that this solution is optimal.

- All demands are routed on single-hop primary paths. Therefore the working capacity cannot be less than m.

- Every edge has a corresponding one-unit demand routed on it, and every vertex is adjacent to at least two edges. Henceforth, every vertex must be incident to at least two protection channels on two distinct edges reserved for protection so that all edge failures can be recovered. Summing over all vertices, and dividing by two (each reserved channel is incident to two vertices), we note that the protection capacity cannot be less than n.

We conclude that there exists a demand D, so that the optimal routing uses all the capacity and the protection-to-working capacity ratio is n/m. This demand set yields the maximum throughput, since it uses all the working capacity and all working paths are one-hop long. ∎

Theorem 3 *Assume that all edges have working capacity Q, and infinite capacity for protection. Then there exists a one-hop demand set D so that the throughput is maximized, and the optimal solution uses* *all the working capacity with the ratio of protection-to-working capacity no less than*

$$\frac{1}{2mQ} \sum_{i \in V} \frac{\delta_i}{\delta_i - 1} Q = \frac{1}{2m} \sum_{i \in V} \frac{\delta_i}{\delta_i - 1}$$

but no more than $n/m = 2/\delta$.

$$\frac{\sum_{i \in V} \frac{\delta_i}{\delta_i - 1} Q}{2mQ} \leq \mathbf{R_s} \leq \frac{2}{\delta} \qquad (11.13)$$

Proof. Consider the demand set of the proof of Theorem 2, multiplied uniformly by a ratio Q, so that the demand between the endpoint of every edge is Q units of capacity. Clearly, a solution consists of routing all demands using single-hop paths along their respective edges, and duplicating the same Hamiltonian circuit $H_c(V, E)$ Q times. This solution has mQ working channels and nQ channels are reserved for protection. This is a valid solution but not necessarily an optimal one. Therefore, the upper bound for the protection-to-working capacity ratio is n/m. The optimal solution uses at least mQ working capacity. This is also the total working capacity available in the network. Using the same argument expressed in Theorem 2, every vertex must have at least $Q\delta_i/(\delta_i - 1)$ protection channels incident to it. Summing over all vertices, and dividing by two, we note that the protection capacity cannot be less than $(1/2) \sum_i Q\delta_i/(\delta_i - 1)$ channels. Hence, the lower bound on the protection-to-working ratio of the theorem is $(1/2mQ) \sum_i Q\delta_i/(\delta_i - 1)$. This lower bound is achievable for the fully connected graph with $(n - 2)$ units of demand between every vertex pair.∎

Note that if $\delta_i = \delta$ (δ being constant), the lower bound becomes $1/(\delta - 1)$, which is the well known lower bound for shared-link or span protection, and is expected in the case of the one-hop demand set used in this theorem.

Theorem 4 *Assume that all edges have working capacity Q, infinite capacity for protection, and that* **all primary paths are at most h hops long**. *Then for any demand set that uses all the network capacity, the protection-to-working capacity ratio is no less than*

$$\frac{1}{2mQ} \sum_{i \in V} \frac{\delta_i}{\delta_i - 1} Q = \frac{1}{2m} \sum_{i \in V} \frac{\delta_i}{\delta_i - 1}$$

but no more than $n(h + 1)/m = 2(h + 1)/\delta$.

$$\frac{\sum_{i \in V} \frac{\delta_i}{\delta_i - 1} Q}{2mQ} \leq \mathbf{R_s} \leq \frac{2(h + 1)}{\delta} \qquad (11.14)$$

Proof. The lower bound is easily verified using the same reasoning as in Theorem 3: the working capacity is no more than mQ and the protection capacity cannot be less than $\delta_i Q/2(\delta_i - 1)$, hence the lower bound. We now prove the upper bound. We assumed that the demand uses all the capacity available, henceforth the working capacity is no less than mQ. A possible solution consists of routing all demands through single-hop paths along their respective edges, and duplicating the same Hamiltonian circuit $H_c(V, E)$ for protection as many times as necessary. Substituting an equivalent number of single-hop demands for every multi-hop demand and using Theorem 2, it is readily shown that $H_c(V, E)$ must be duplicated at least Q times. However, because demands may use multi-hop routes, and thus overlap, the protection circuit must be further duplicated in order to avoid sharing violations. If a demand is routed along an h-hop path, then this demand belongs to a set of at most $1 + (Q - 1)h \leq Q(h + 1)$ mutually interfering demands, i.e. demands whose primary paths have one or more edges in common. Therefore, $H_c(V, E)$ only needs to be duplicated at most $Q(h + 1)$ times. We conclude that for all demand sets occupying all of the network capacity, there is a feasible solution which uses mQ working channels, and $n(h + 1)Q$ protection channels. Hence, the working to protection ratio $n(h + 1)/m$. These bounds are more general and broader than the bounds found in Theorem 2. Ring topologies, with any demand connectivity that requires Q working channels per edge, are examples of networks that achieve the lower bound.■

Theorem 5 *Assume that all edges have nonnull, nonuniform working capacity in the range* $[Q_{\min}, Q_{\max}]$, *infinite capacity for protection, and that* **all primary paths are at most** h **hops long**. *Then, for any demand set that uses all the network capacity, the protection-to-working capacity ratio is no less than*

$$\frac{1}{2mQ_{\max}} \sum_{i \in V} \frac{\delta_i}{\delta_i - 1} Q_{\min} = \frac{Q_{\min}}{2mQ_{\max}} \sum_{i \in V} \frac{\delta_i}{\delta_i - 1},$$

but no more than

$$\frac{Q_{\max}}{Q_{\min}} \frac{n(h + 1)}{m}.$$

$$\frac{Q_{\min}}{Q_{\max}} \frac{\sum_{i \in V} \frac{\delta_i}{\delta_i - 1}}{2m} \leq \mathbf{R_s} \leq \frac{Q_{\max}}{Q_{\min}} \frac{2(h + 1)}{\delta} \qquad (11.15)$$

Proof. We repeat the proof of Theorem 4, and appropriately choose the edge capacities in the interval $[Q_{\min}, Q_{\max}]$ so that we obtain the lower and upper bounds shown in Equation 11.15.■

The main convergence theorem (Theorem 1) is a direct consequence of Theorem 5.

11.5 Dimensioning Mesh Optical Networks

In this section, we introduce a node model and apply a set of traffic conservation equations to derive certain key network quantities, namely the total demand that can be supported on any given link, node, or across a protected mesh network [183, 188].

11.5.1 Node Model and Traffic Conservation Equations

We introduce the set of parameters shown in Table 11.1. We model a node as shown in Figure 11.11. The ports on a switch are categorized as either add/drop ports that are facing toward the outside of

Table 11.1: Dimensioning parameters

S	size of switch
N_a	number of add/drop ports
N_{ot}	number of network-side ports used by originating/terminating working path
N_{th}	number of network-side ports used by through working path[3]
N_p	number of protection ports (DBPP and SBPP cases)
γ	switch loading
P_r	ratio of add/drop ports used for LAPS protection[4] to add/drop ports used for service[5]
T	ratio of through to working capacity[6]
$\mathbf{R}$	ratio of protection to working capacity

---

[3]Working path that is not terminating, but rather it traverses the node.

[4]LAPS stands for Linear Automatic Protection Switching and refers to the protection provided at the drop side of a switch.

[5]Drop-side protection refers to ports on the drop side of a switch (as opposed to the network side) that are dedicated to provide protection to working drop-side ports. P_r is 0 if no drop-side protection is used; 1 if $1 + 1$ drop-side protection is used; $1/N$ if $1 : N$ drop-side protection is used.

[6]Working capacity includes ports used for through working paths.

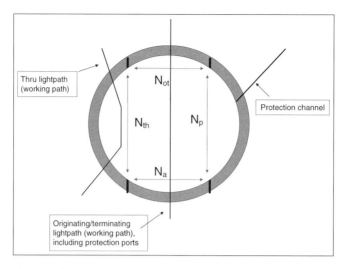

Figure 11.11: Node model for traffic conservation. (From [190], Figure 10. Reproduced by permission of © 2005 The Institute of Electrical and Electronics Engineers.)

the network and connected to client equipment, or network-side ports, that are facing towards the inside of the network and support trunks connecting the nodes to each other. The primary path of an originating/terminating lightpath uses one or more add/drop ports from the pool of N_a ports and a network-side port from the pool of N_{ot} ports. The primary path of a through lightpath uses two ports from the pool of N_{th} ports. A protection channel on the backup path of either an SBPP or a DBPP lightpath uses a port from the pool of N_p ports. The sizes of the pools obey the following conservation equations:

$$N_a + N_{ot} + N_{th} + N_p = \gamma S \tag{11.16}$$

$$N_a = N_{ot}(1 + P_r) \tag{11.17}$$

$$N_{th} = T(N_{ot} + N_{th}) \tag{11.18}$$

$$N_p = \mathbf{R}(N_{ot} + N_{th}) \tag{11.19}$$

Equation 11.17 captures the fact that some of the drop-side ports are used for drop-side protection.

Given a path of length h, the path traverses $2(h-1)$ ports at through or intermediate switches (two per switch) while the path uses two additional ports on the network side of the originating and terminating switches, yielding $T = 2(h-1)/2h = (h-1)/h$. Rewriting as $1 - T = 1/h$, and plugging Equations 11.17, 11.19, and 11.18 into Equation 11.16, we obtain after simplification:

$$N_a = \gamma S \frac{1 + P_r}{1 + P_r + (1 + \mathbf{R})h} \tag{11.20}$$

In the case of lightpaths with no drop-side protection ($P_r = 0$) and no network-side protection ($R = 0$), we have $N_a = \gamma S/(1 + h)$, as expected. The average number of lightpaths in a maximally loaded mesh network (in both DBPP and SBPP cases) can then be derived as

$$L_{\text{network}} = \frac{N_a \times n}{2(1 + P_r)} = \frac{\gamma S}{2} \frac{n}{1 + P_r + (1 + \mathbf{R})h} \tag{11.21}$$

Note that $L_{network} \propto O(n/h) = O(n/\ln n)$. From Equation 11.21, and using $\delta = 2m/n$, we can write the average number of lightpaths per link in a maximally loaded mesh network as

$$L_{link} = \frac{L_{network} \times h}{m} = \frac{\gamma S}{\delta} \frac{h}{1 + P_r + (1 + \mathbf{R})h} \tag{11.22}$$

Note that $L_{link} \to \gamma S/[\delta(1 + \mathbf{R})]$ when $n \to \infty$, independent of h. The average number of lightpaths per node in a maximally loaded mesh network is

$$L_{node} = \frac{L_{network} \times (1 + h)}{n} = \frac{\gamma S}{2} \frac{1 + h}{1 + P_r + (1 + \mathbf{R})h} \tag{11.23}$$

Note that $L_{node} \to \gamma S/[2(1 + \mathbf{R})]$ when $n \to \infty$, independent of h. The formula for the number of lightpaths in a maximally loaded network is a function of the protection ratio $\mathbf{R}$. $\mathbf{R}$, in the case of SBPP lightpaths, depends in turn on the number of lightpaths in the network, through the sharing factor of shared backup channels, F. Therefore, determining $L_{network}$ and $\mathbf{R_s}$ requires solving a fixed point equation.

11.5.2 Dimensioning Examples and Results

Let us demonstrate how these formulas can be used to dimension optical mesh networks. For reasonable size networks, the authors in [190] have measured $\mathbf{R}$ to be in the range of $1.2-1.5$ for Dedicated Backup Path Protection and $0.4-0.8$ for Shared Backup Path Protection. Also, operational networks are usually run at around 70% utilization. Finally, we assume here that $P_r = 0$. Figure 11.12 plots the maximum number of lightpaths as a function of the number of nodes (with average node degree 3) for the case of unprotected demand ($\mathbf{R} = 0$), SBPP demands ($\mathbf{R} = 0.7$), and DBPP demands ($\mathbf{R}$ obtained from Equation 11.10). Two sets of curves are given for switches of size 512 ports, with $\gamma = 0.7$ and $\gamma = 0.9$ utilization levels. From these curves, it is easy to determine the maximum number of lightpaths that can be supported for a given network size and at a given network utilization. Inversely, given a certain amount of traffic that needs to be carried, it is easy to estimate the number of nodes given other characteristics such as average node degree and switch size.

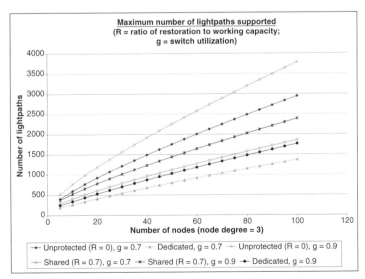

Figure 11.12: Maximum number of lightpaths as a function of number of nodes for different switch utilizations γ (switch size = 512). (From [190], Figure 11. Reproduced by permission of © 2005 The Institute of Electrical and Electronics Engineers.)

11.6   The Network Global Expectation Model

Network analysis is not only used by network designers and operators, but also by other members of the value chain. From end users at one extreme, to equipment designers and enabling technology researchers at the other, being able to relate network needs to network element requirements quickly and with clarity is very desirable. However, the uniqueness and complexity of networks often obscure common characteristics, scaling behavior, and trends. Analytical models of networks can serve to bridge the gap between the detailed analysis of a particular network and the general behavior of classes of related networks and traffic demand.

As an example, Korotky, Bhardwaj and their collaborators have focussed on developing comprehensive, yet tractable, high-level analytical descriptions of network requirements, such as capacity dimensioning [41, 42, 43, 184, 185]. Their approach, referred to as the Network Global Expectation Model because it addresses statistical expectations for the entire network, begins with the concept that a key aspect in the comparison and selection of network architectures and their technological implementations is the total cost of ownership of the network. And while operational and management expenses represent a large portion of the total cost of ownership, capital costs are recognized as a considerable and highly visible part of the initial investment and, consequently, are a significant factor in the choice of architecture and technology. The intent of their work is to provide a versatile means to quickly gauge the network equipment needs and costs using only modest computational resources. As this objective is well aligned with the goals of this book and specifically this chapter, we include in this section a brief overview of the main concepts, formalisms, results, and applications of the Network Global Expectation Model.

In the Network Global Expectation Model (NetGEM), expectation values of network variables and functions of these variables are formally evaluated by averaging over the entire network to establish either exact or approximate analytical relationships between dependent and independent variables. Through this approach the global (network) and local (network element) views of the communications system are naturally and accurately connected, and results for a very wide range of network sizes and large number of variations can be computed very fast with useful accuracy. As the cost of the network for a specified set of features is considered the metric for comparison of architectures and technologies, the NetGEM is constructed from this perspective.

The telecommunications network is represented by the combination of a network graph, denoted G, consisting of a set of n nodes and a set of m connecting links, or edges, and a given network traffic demand. The network traffic demand is represented by the symmetric demand matrix T, with elements t_{ij}, the total number of two-way demands D, and the total ingress/egress traffic $T = \sum_{ij} t_{ij}$. The primary model input variables are taken to be $G(n, m)$, D, and T together with the demand model. All other variables of interest may be determined from these. For specificity, individual demands are considered to be carried on an individual channel, e.g., a wavelength.

For consistency with the authors' original publication, we have tried to use the same notations that they used. Specifically, the average of a set is represented by angle brackets, i.e. the average of a set s is $\langle s \rangle$, and the covariance of two variables, p and q, is $\sigma^2(p, q)$. At the same time, we have changed some of the variable names to be consistent with the naming conventions used throughout this chapter. In particular, the number of wavelengths or channels used for capacity, either working capacity to carry demands or protection capacity to recover traffic in case of a failure, is denoted by the variable W with the subindex P, as W_P, to specify protection capacity. The ratio of protection capacity to working capacity is denoted by R (vs κ by the authors), again for consistency with the remainder of this chapter.

As summarized earlier, the average degree of a node $\langle \delta \rangle$ is calculated by summing the number of links and by dividing by the number of nodes yielding $\langle \delta \rangle = 2m/n$. The number of hops between a pair of terminals is defined as the minimum number of links traversed by a demand between the terminating node-pair. The expectation value of the number of hops is over the set of demands. For minimum hop routing of uniform demand over two-dimensional networks the authors have found that $\langle h \rangle$ is accurately approximated by

$$\langle h \rangle = \sqrt{\frac{n-2}{\langle \delta \rangle - 1}} \qquad (11.24)$$

The mean number of channels, $\langle W \rangle$, appearing on a link of the network is expressed as

$$\langle W \rangle = \frac{\langle d \rangle \langle h \rangle}{\langle \delta \rangle} \qquad (11.25)$$

where $\langle d \rangle$ is the mean number of demands terminating at a node. This important result is exact and valid independent of the demand model. Note, however, that the value of $\langle h \rangle$ is implicitly dependent upon the demand model, as specified above.

Extra capacity for protection is expressed as a fractional increase, $\langle R \rangle$, in the total deployed capacity to service the traffic and provide survivability relative to the case without survivability and using minimum hop routing. The average number of channels on a link, including extra capacity for protection, is expressed as $\langle W \rangle + \langle W_P \rangle = \langle W \rangle (1 + \langle R \rangle)$. Later in this section we summarize the statistical estimates of the extra capacity, $\langle W_P \rangle$.

The mean number of ports required on a cross-connect to service the working and network-side protection channels present at a node is $\langle d \rangle + \langle W \rangle (1 + \langle R \rangle) \langle \delta \rangle$, which is observed to scale approximately as $n^{3/2}$ for large n. Independent of the demand model, the ratio of the number of terminated channels to total (termination and through) channels present at a node is given by $2/1 + \langle h \rangle$, which has been found to scale as $2/\sqrt{n}$ for large n. The mean length of a link is estimated given the geographic area, A, serviced by the network using the approximation $s \approx \sqrt{A}/(\sqrt{n} - 1)$.

By assuming a cost structure for the network elements, the above relationships can be used to compute costs for categories of network elements, as well as total network costs [183].

Carrier class networks must provide high availability and so must be survivable against the failure of subsystems. One measure of the efficiency of the network design is the amount of extra capacity that must be deployed to ensure complete survivability against any single link failure. A link failure results in the failure of all demands on the failed link, which are then rerouted over backup paths utilizing the protection capacity in the network. Bhardwaj et al. [183] derive the protection capacity in general mesh networks by decomposing the W demands on a link into the terminating demands, $\langle W^t \rangle$, i.e. demands that terminate at one of the nodes attached to the failed link, and through demands, $\langle W^{th} \rangle$. Their strategy is to consider the extra capacity requirements for $\langle W^t \rangle$ and denote it as $\langle W_P{}^t \rangle$. Later, they consider the incremental extra capacity required on a link to reroute demands not terminating at the adjacent node, the through demands $\langle W^{th} \rangle$, and denote this incremental extra capacity as $\langle W_P{}^{th} \rangle$. The total average extra capacity on a link, $\langle W_P \rangle$, is then the sum of $\langle W_P{}^t \rangle$ and $\langle W_P{}^{th} \rangle$. The demands on a link are counted relative to one of the nodes attached to the link. If d_i are the terminating demands at a node n_i of degree δ_i, then on average there are d_i/δ_i terminating demands on each link connected to n_i and thus using Equation 11.25, $\langle W^t \rangle$ and $\langle W^{th} \rangle$ can be approximated by Equations 11.26 and 11.27, respectively.

$$\langle W^t \rangle = \frac{\langle W \rangle}{\langle h \rangle} \qquad (11.26)$$

$$\langle W^{th} \rangle = \langle W \rangle - \langle W^t \rangle \approx \langle W \rangle \left(1 - \frac{1}{\langle h \rangle}\right) \qquad (11.27)$$

Consider the links terminating at node n_i, which are δ in number and denote the terminating working capacity and the protection capacity on any of those links (e.g., link m_{ij}) as W_{ij}^t and $W_{P_{ij}}^t$, respectively. If a particular link fails, then the sum of the protection capacities of the surviving links connected to node n_k must be greater than or equal to the failed terminating demands on link m_{ij}

to be able to recover the terminating traffic, i.e. $\sum_{j\neq k}^{\delta_i} W_{P_{ij}}^t \geq W_{ik}^t$. This equation represents the local bound on the protection capacity at a node. With this equation as the starting point, the global average of the terminating protection capacity can be derived as

$$\langle R^t \rangle = \frac{\langle W_P^t \rangle}{\langle W^t \rangle} \geq \left(\frac{1}{\langle h \rangle \, (\langle \delta \rangle - 1)} + \frac{\sigma^2(\delta, W_P^t) - \sigma^2(\delta^2, W_P^t)}{\langle d \rangle \, \langle h \rangle \, (\langle \delta \rangle - 1)} \right) \left(\frac{1}{1 + \sigma^2(\delta)/\langle \delta \rangle \, (\langle \delta \rangle - 1)} \right) \quad (11.28)$$

The dominant term on the right-hand side of Equation 11.28 is the first term from the left within the first parentheses. It has a contribution from the ratio of $\langle W^t \rangle$ and $\langle W \rangle$ in the form of $1/\langle h \rangle$ (see Equation 11.26) and a contribution from the ratio of the failed to surviving links connected to the adjacent node in the form of $1/(\langle \delta \rangle - 1)$. The term in the second parentheses from the left is a multiplication factor that incorporates the effects of topological variations. The derivation of Equation 11.28 assumed that all surviving links connected to the node participate in recovering the terminating demands on the failed link, i.e. there are $\langle \delta \rangle - 1$ disjoint backup paths available. This therefore represents the maximal sharing of protection capacity and consequently the minimum extra capacity requirement and is referred to as the *divisible bound* of the extra capacity. At the other extreme, only one backup path might be found, and in this scenario two links attached to a node must be assigned protection capacity equal to $\langle W^t \rangle$. Thus, in this case the total protection capacity on all links attached to the node is on average $2\langle W^t \rangle$, and the total working capacity is $\langle \delta \rangle \langle W \rangle$. The dominant term contributing to $\langle R^t \rangle$ is then $2\langle W^t \rangle / \langle \delta \rangle \langle W \rangle = 2/\langle h \rangle \langle \delta \rangle$. In this scenario the constraint expressed by the summation example at a single node is modified so that the summation on the left-hand side of the inequality reduces to one term, i.e. the link chosen for protection must have protection capacity greater than the failed terminating demand. The global expectation of the terminating extra capacity, $\langle R^t \rangle$, for the limiting condition where there is only a single backup path, referred to as the *indivisible bound*, is expressed as

$$\langle R^t \rangle = \frac{\langle W_P^t \rangle}{\langle W^t \rangle} \geq \left(\frac{2}{\langle h \rangle \, \langle \delta \rangle} - \frac{\sigma^2(\delta^2, W_P^t)}{\langle d \rangle \, \langle h \rangle \, \langle \delta \rangle} \right) \left(\frac{1}{1 + \sigma^2(\delta)/\langle \delta \rangle^2} \right) \quad (11.29)$$

Since the average of the number of protection paths lies between 1 and $\langle \delta \rangle - 1$, the optimum value of $\langle R^t \rangle$ lies between the divisible and indivisible bounds.

Having derived the bounds on the terminating extra capacity W_P^t, the following analysis focuses on the through extra capacity requirement $\langle W_P^{th} \rangle$. When a link fails in the network, the failed demands–on average $\langle W \rangle$ in number–are routed over protection paths of average length denoted as $\langle h' \rangle$. The protection path of a failed demand may be selected from among $m - \langle h \rangle$ links, as the protection path is disjoint from the working path. Of the $\langle W \rangle$ failed demands, the terminating $\langle W^t \rangle$ demands can be recovered by the assignment of $\langle W_P^t \rangle$ extra capacity to all links in the network as derived above. The maximum additional average protection capacity required on all links in the network to recover the $\langle W^{th} \rangle$ through demands is denoted by $\langle W_P^{th} \rangle_{max}$. As the total additional protection capacity required is $\langle W^{th} \rangle \langle h' \rangle$, $\langle W_P^{th} \rangle_{max}$ can be expressed as

$$\langle W_P^{th} \rangle_{max} = \frac{\langle W^{th} \rangle \langle h' \rangle}{m - \langle h \rangle} = \langle W \rangle \left(1 - \frac{1}{\langle h \rangle} \right) \left(\frac{\langle h' \rangle}{m - \langle h \rangle} \right) \quad (11.30)$$

and the maximum contribution to $\langle R \rangle$ by the through demands, $\langle R^{th} \rangle_{max} = \langle W_P^{th} \rangle_{max} / \langle W \rangle$, can be expressed as

$$\langle R^{th} \rangle_{max} = \left(1 - \frac{1}{\langle h \rangle} \right) \left(\frac{\langle h' \rangle}{m - \langle h \rangle} \right) \quad (11.31)$$

The analytical model of extra capacity can be applied to different failure recovery strategies, e.g. SBPP, link protection, by the proper choice of $\langle h \rangle$ and $\langle h' \rangle$. For Shared Backup Path Protection, $\langle h \rangle$ is

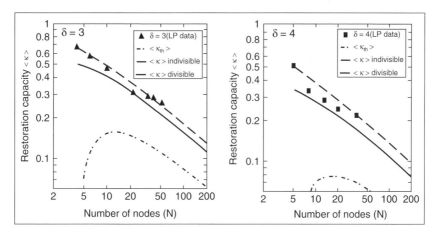

Figure 11.13: Extra capacity for SBPP in regular planar networks of (a) degree 3 and (b) degree 4.

the average length of the working paths, and $\langle h' \rangle$ is the average length of the shared backup paths. For nominally two-dimensional networks, $\langle h \rangle$ may be approximated semiempirically by Equation 11.24 and $\langle h' \rangle$ may be approximated semiempirically as being equal to twice $\langle h \rangle$. For networks having a more linear character, i.e. networks with a significant number of degree 2 nodes resulting in a mean degree of less than 3, the relationship between $\langle h \rangle$ and n becomes more linear and both the mean working and backup path lengths increase. Note that, in the limit of a large bidirectional ring network with location-independent demand profile, $\langle h \rangle$ scales as n and $\langle h' \rangle \approx 3 \langle h \rangle$. Figure 11.13 illustrates the comparison of the extra capacity calculated using the NetGEM model and that calculated using LP simulations for regular mesh networks with uniform demand utilizing Shared Backup Path Protection. Readers interested in link protection, and how this model can be applied, are referred to [183].

11.7   Accuracy of Analytical Estimates

Recently, work by Forst and Grover in [123] presented an extensive analysis of the analytical estimates introduced earlier in this chapter and published in [183] and [188, 190]. In their work, the authors of [123] conducted a further series of tests of the equations to assess their accuracy over a more general and extensive set of network topologies and demands than the ones used to develop the models. To carry out this assessment, the authors in [123] compared the results from the analytical models to solutions obtained by solving ILP-based formulations of the network designs. It was found that the mathematical models of [183] and [188, 190] could have typical errors of up to 30%. In addition, the authors in [123] provided a number of insights into network-dependent phenomena and the effects of network topology (nodal degree, hop and distance characteristics) and traffic demand (variance, proximity) on the design of SBPP networks. This points out the need for further research to determine whether analytical models can be developed that perform well over all realistic network and demand models.

11.8   Recovery Time Performance

Recovery times in Shared Backup Path Protected networks can be analyzed by decomposing a recovery event into the different recovery phases: the failure detection and notification phase, the recovery signaling phase, and the traffic reestablishment phase [43, 190].

The recovery architecture implemented in a network will have a key impact on the recovery times. For example, the mechanism used for detection will drive the detection time (e.g. IP layer handshake vs SONET layer detection). The mechanisms used for notification and signaling will also drive the recovery time (e.g. out-of-band TCP/IP-based notification and signaling vs in-band per-lightpath signaling using SONET overhead bytes). The second impact comes from the internal switch architecture and the queueing that results when processing detection, notification, and signaling messages during a recovery event. In addition, the traffic loading in the network will drive the queueing mechanisms invoked during a recovery event, and will also directly impact on the recovery times.

The average failure detection and notification time T_{DN} can be expressed as follows:

$$T_{DN} = T_D + \frac{h}{2} \times f(T_P) + \frac{h}{2} \times T_{1\text{hop}} \qquad (11.32)$$

where T_D is the failure detection time (for example, 10 ms for SONET-layer detection), h is the average length of the primary path (in number of hops), T_P is the processing time of an event (e.g. a failure detection or failure notification event) at each node, $f(.)$ is the queueing function for the processing time associated with an event (detection, notification, or recovery signaling), and $T_{1\text{hop}}$ is the average propagation time on a network link.

The average recovery signaling time can be expressed as follows:

$$T_S = (h'' + 1)(f(T_P) + g(T_{XC})) + h'' \times T_{1\text{hop}} \qquad (11.33)$$

where h'' is the average length of the backup path (in number of hops), T_{XC} is the cross-connect time at a node, and $g(.)$ is the queueing function for the processing time of cross-connect requests to the switching fabric.

The recovery architecture as well as the internal node architecture can both significantly influence the overall recovery times. For example, per-demand failure notification and signaling using the overhead bytes of the SONET signals carrying the demands creates a faster and more robust architecture than one where all the notification and recovery signaling is carried over an out-of-band TCP/IP network [38]. Also, an internal switch architecture could be implemented that allows cross-connect requests to be grouped and processed in parallel, thereby improving the cross-connect time component of the recovery process [94, 169]. Finally, the loading of the network will directly drive the queuing behavior as a large amount of traffic can be impacted during a failure, and therefore more recovery requests have to be handled.

As previously discussed in Chapter 4 (Section 4.5), SBPP studies using simulation tools have shown that recovery times for the Tellium's STS-48 switch were mainly influenced by the number of failed lightpaths processed by a switch during recovery [16, 18]. In particular, the worst case occurred when all lightpaths terminate at the same two end-switches rather than at switches distributed throughout the network. Furthermore, simulation studies for the Tellium's STS-48 switch have shown that, for a given topology and a given set of primary and backup routes, the recovery time increased roughly linearly with the number of lightpaths that failed simultaneously [16, 18]. Thus, a coarse analytical approximation was constructed which assumed the worst-case scenario involving the maximum number of lightpaths that are processed by the same end-nodes [190].

Figure 11.14 from [190] compares the coarse analytical results obtained for a 50-node network with the simulation results, on a y-axis logarithmic scale. This figure also appears in Chapter 4 (Figure 4.20) but is repeated here for the reader's benefit. It was assumed that all those lightpaths originate and terminate at the same end-nodes. Therefore, the recovery requests contend for resources at the end-nodes to perform bridge and switch cross-connects. These are very conservative assumptions in a worst-case scenario. Most likely, under more realistic assumptions, one would observe shorter recovery times. Having validated the basic model and parameters, the authors used their approximation formulas for different networks and estimated the recovery times one could expect for a maximally

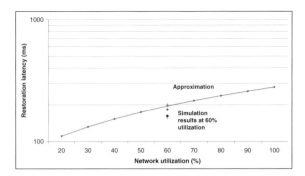

Figure 11.14: Analytical vs simulation results for hypothetical 50-node network. (From [190], Figure 12. Reproduced by permission of © 2005 The Institute of Electrical and Electronics Engineers.)

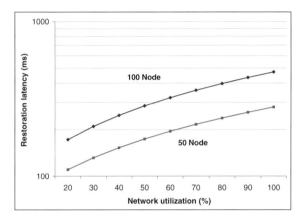

Figure 11.15: Recovery latency as a function of the total network utilization γ. (From [190], Figure 14. Reproduced by permission of © 2005 The Institute of Electrical and Electronics Engineers.)

loaded network at different utilization levels. The analytically calculated recovery latency curve versus the network utilization is shown in Figure 11.15 (again on a y-axis logarithmic scale). The authors in [190] conducted a study of a real 50-node network with a utilization of 60% (where $L = 36$ lightpaths failed for the analytical approximation) and a backup channel sharing ratio of 0.46. In Figure 11.14, the authors superimpose the simulation results for five single failure events affecting the most number of lightpaths at 60% utilization. As can be seen from this figure, the analytical approximation yields a recovery latency which is within the same order of magnitude of the results obtained using simulation. This behavior is typical of similar studies we have performed for different networks.

Certainly, the results presented here are specific to the recovery architecture that was used with Tellium's switches, and we expect the performance results to be very sensitive to changes in the architecture, including the internal switch software and hardware design and implementation.

11.9 Conclusion

In this chapter, we presented the models and corresponding collections of approximation formulas proposed by Korotky and Bhardwaj [183] and Labourdette *et al.* [190] that quickly but roughly estimate a network size and failure-recovery performance with limited inputs. In particular, we presented sets of equations that relate number of sites, average fiber connectivity, demand load and capacity for various failure-recovery architectures, with idealized network and demand models. These results can be used to easily and quickly estimate the amount of traffic that can be carried over a given network, or, inversely, given the traffic to be supported, to assess the characteristics of the topology required (in terms of number of nodes, connectivity). Finally, this approach can be used to estimate the recovery performance that can be expected from a network without requiring any extensive simulation studies. As shown recently by Forst and Grover [123] in a series of tests to assess accuracy over a more general and extensive set of network topologies and demands than the ones considered to develop and use the models, those purely analytical models have limitations compared to formal optimization and simulation methods. This points out the need for further research to determine whether analytical models can be developed that perform well over all realistic network and demand models.

Chapter 12

Service Availability in Path-Protected Mesh Networks

12.1 Introduction

Network service availability is a critical element of a service-level agreement and is typically evaluated based on the number of unavailable minutes per year. A widespread belief is that networks with faster recovery times provide higher service availability, created by the assumption that fast recovery from a failure leads to smaller downtime. This is not necessarily the case and it has been shown that availability has in fact little to do with the recovery speed when the recovery time is small compared to the time needed to replace failed network elements or to repair fiber cuts.

In this chapter, we propose a model for analyzing the service availability of Dedicated Backup Path-Protected (DBPP) and Shared Backup Path-Protected (SBPP) services. Within that model, we discuss the impact on service availability of two additional recovery schemes, channel protection and reprovisioning. We then compare the availability in path-protected mesh networks using a multi-domain recovery approach with that of the same networks using a single-domain recovery approach. Finally, we compare the availability of path-protected mesh networks with that of span-protected ring-based networks.

This chapter is organized as follows. In Section 12.2, we introduce the concept of service availability, and give a brief introduction of reliability and availability calculations. In Section 12.3 we present a Markov model for the sequence of events (failures, recoveries, repairs) that lead to an outage and to the service being unavailable. We discuss how this model applies to path protection within a single recovery domain, while considering additional recovery schemes such as channel protection and reprovisioning. In Section 12.4 we discuss path protection across single and multiple recovery domains, and show how splitting the network into multiple domains increases the overall availability. Section 12.5 compares service unavailability in path-protected mesh networks and more traditional networks of interconnected rings. We conclude in Section 12.6.

12.2 Network Service Availability

12.2.1 Motivation

While fast recovery (of the order of say, 100 ms) is an aspect of service availability, much more critical yet is the ability to recover from multiple failures quickly as opposed to the several hours it

Path Routing in Mesh Optical Networks   Eric Bouillet, Georgios Ellinas,
Jean-François Labourdette, Ramu Ramamurthy   © 2007 John Wiley & Sons, Ltd

takes in practice to replace damaged equipment. A number of authors have studied service availability in path-protected mesh networks [79, 81, 95, 98, 141, 169, 300], span-protected networks [77], as well as *p*-cycle protected networks [80, 271]. One of the key factors improving service availability is the ability to recover from multiple failures, with, for example, a combination of protection for the first failure, and reprovisioning for any subsequent failure(s).

Indeed, the availability of a network has very little to do with the recovery speed, provided the recovery speed is small compared to Mean Time to Repair (MTTR), which is typically several hours. The argument was made originally and relatively recently by Grover and his team and supported with numerical demonstrations in [77, 79, 96], as well as Sengupta *et al.* [275] in the context of comparing availability in ring-protected and mesh-protected networks.

The availability of a network is measured in terms of unavailable minutes per year. For example, a network with availability greater than 99.999% implies that a circuit on average must not be out of service for more than 5.26 minutes per year. If we assume that a circuit would fail once a year and is recovered within 60 ms, the availability would be 99.9999998%. On the other hand, if the recovery speed was 200 ms, the availability would be 99.9999994%. This simple calculation clearly demonstrates that recovery speed has very little to do with availability. The primary contributor to the unavailability of a protected circuit arises from the fact that when a circuit needs to be recovered, the recovery capacity may not be available due to other prior failures not yet repaired. That would happen if the protection path fails before the failed primary path is repaired. It could also happen if a shared backup channel is occupied and cannot protect another lightpath whose primary path has failed. In order to analyze service availability, it is necessary to develop quantitative models for both mesh and ring architectures and compute the probability of a circuit's unavailability using those models. In this chapter, we propose some models and compare the availability in single and multi-domain recovery mesh networks and of ring and mesh network architectures.

12.2.2   Focus on Dual-Failure Scenarios

Mesh networks supporting Dedicated Backup or Shared Backup Path-Protected lightpaths are immune to single failures (unless it is the failure of the originating or terminating node for a given lightpath). Thus, service availability becomes of interest in the context of dual-failure scenarios. Indeed, many operational networks are protected against single failures and the domain of concern and interest of both network operators and end-customers of network services is the tolerance to double failures. Because network maintenance and growth activities (including network upgrades) require particular spans to be taken out of service, networks are often in a state equivalent to single-failure scenarios, in which the occurrence of a single failure would be equivalent to the occurrence of dual failures [98, 143]. Additionally, the presence of known or unknown shared risk groups (node, link, or equipment) can turn a single failure into the equivalent of a dual-failure scenario. However, the intent when designing networks is not to protect against an actual dual-failure scenario, which would be cost-prohibitive, but to understand them and to minimize both their occurrence and their impact where possible. For further discussions on motivations for dual-failure considerations, see [96]. Service-Level Agreements (SLAs) are frequently demanded by customers (as part of Request For Proposals (RFPs)) and are typically addressed by service providers by analyzing the risk that service-affecting failures will occur and by assessing the financial penalty that would result.

12.2.3   Reliability and Availability

Let us introduce some basic definitions for the reliability and availability of systems and describe how they apply in the context of telecommunications networks. The treatment provided in this section is concise and only intended to introduce the key concepts of reliability and availability as well as some basic formalism to compare service availability in different path-protected mesh networks and ring-protected networks. For a more complete treatment of this subject, the reader is referred to Chapters 3 and 8 in the book by Grover [139].

The *reliability* of a system is *the probability that the system will perform as intended over a specified period of time.* Let us introduce a few terms that are commonly used to describe reliability. The Mean Time To Failure (MTTF) of a system is the average time from an operational state to the next failure, and is the inverse of the failure rate λ.

$$\lambda = \frac{1}{\text{MTTF}} \tag{12.1}$$

Of particular interest in telecommunications networks is the reliability of the fiber infrastructure. Failure frequency associated with fiber cuts depends on the geographical location of the fiber, and as one would expect, is higher in metropolitan areas than in rural areas. Typical numbers range from a few cuts per year per 1000 miles of fiber in the long-haul portion of the network, to more than 10 cuts per year per 1000 miles in metropolitan areas. For a more complete treatment and figures on such failure rates, see Chapter 3 of the book by Grover [139].

The Failure In Time or FIT rate is a standard unit used for specifying failure rates, or conversely MTTFs, and is defined as the failure rate in 10^9, or 1 billion, hours. It is defined by equipment vendors for the different components and cards that make telecommunications switches, as well as generically defined by Telcordia Technologies standards [2, 3, 5, 6, 7].

$$\text{FIT} = \frac{10^9}{\text{MTTF}} \tag{12.2}$$

The Mean Time To Repair (MTTR) for a system is the average time to repair a failure, and is the inverse of the repair rate μ.

$$\mu = \frac{1}{\text{MTTR}} \tag{12.3}$$

Typically, the MTTR for network elements (e.g. a card) is the time it takes to dispatch a technician to a Central Office (CO) or a customer location to replace the failed component. Service providers usually offer SLAs for repairing failed network elements that can range from two to four hours. In the case of a fiber cut, the MTTR can be higher because the incident may have occurred in an isolated area and the time required to localize the cut and dispatch a technician increases as a consequence. Furthermore, work such as fiber splicing may be required, which adds to the repair time. Overall, an MTTR of eight hours for a fiber cut would be reasonable. In the case of submarine cables, the difficulty of accessing the cable yields MTTRs that can range from 24 to 48 hours.

The Mean Time Between Failures (MTBF) of a repairable system is the average time between two subsequent failures, and is strictly the sum of the MTTF and the MTTR, thus including the time to repair following a failure. Because the MTTR is typically small compared to the MTTF (MTTR $\ll$ MTTF), the MTBF is often equal to the MTTF.

$$\text{MTBF} = \text{MTT}F + \text{MTT}R \simeq \text{MTTF} \tag{12.4}$$

The *availability* of a system is *the fraction of time that the system is performing as intended over a period of time.* Availability is different from reliability for a repairable system because it covers the operation of a system over sequences of failure-free periods and repair periods. Note that the availability of a system can be high even if individual components have low reliability, if failures are repaired quickly and/or if there is enough redundancy built into the design for the system to continue operating after a component failure. Unavailability in telecommunications networks is typically measured in minutes per year. Typical availability objectives for critical telecommunications services would be at least 99.999% (so-called five nines or 5 $9's$), which represents about 5.26 minutes of downtime per year. The availability A of a system, and its converse, the unavailability $U \equiv 1 - A$ of a system, are expressed as:

$$A \quad = \quad \frac{\text{MTTF}}{\text{MTTF} + \text{MTTR}} = \frac{\text{MTTF}}{\text{MTBF}} = \frac{\mu}{\lambda + \mu} \tag{12.5}$$

$$U \quad \equiv \quad 1 - A = \frac{\text{MTTR}}{\text{MTTF} + \text{MTTR}} = \frac{\lambda}{\lambda + \mu} \tag{12.6}$$

12.2.3.1   Availability of Networks of Components

Networks are built of individual network components assembled in parallel, series, and combination thereof. In a parallel configuration as shown in Figure 12.1(a), the network remains operational as long as at least one of the components is operational. For example, in a Dedicated Backup Path-Protected configuration, the lightpath remains operational as long as either the primary or backup path remains operational. Assuming that the components fail independently and each has unavailability U_i, the unavailability U_p of a network of parallel components is expressed as the product of the individual component unavailabilities U_i:

$$U_p = \prod_i U_i \tag{12.7}$$

In a serial configuration as shown in Figure 12.1(b), the failure of any of the serial components causes the network to fail. For example, the failure of any component on a DWDM link (DWDM equipment, optical amplifier, fiber cut, etc.) causes that link to fail. The availability A_s of a network of serial components is expressed as the product of the individual component availability $A_i = 1 - U_i$ as:

$$A_s = \prod_i A_i \tag{12.8}$$

and the unavailability U_s can be approximated as the sum of the individual component unavailabilities U_i, as long as $U_i \ll 1$:

$$U_s = 1 - A_s = 1 - \prod_i (1 - U_i) \approx \sum_i U_i \tag{12.9}$$

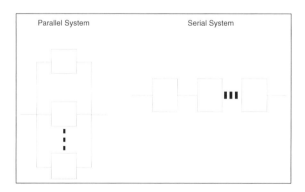

Figure 12.1: Networks of (a) parallel components and (b) serial components.

12.3 Service Availability in Path-Protected Mesh Networks

The primary path of a protected lightpath is protected by a node and/or link-disjoint backup path through dedicated backup or shared backup path protection. Service unavailability would therefore only occur as a result of multiple concurrent failure scenarios, and those failures would be typically caused by a fiber cut or equipment (WDM, amplifier, transceiver) failure. We propose Markov models based on the sequences of events from working state to service outage to model and compute the service availability of path-protected lightpaths.

12.3.1 Dual-Failure Recoverability

A comprehensive approach to determining the availability of a particular service or lightpath consists of the following as described in [96]. One first determines the dual-failure recoverability $R(a, b)$ of the affected service paths for each dual span failure[1] of spans a and b. The dual-failure recoverability counts the ratio of recovered to affected paths for the dual span failure (a, b). Then, knowing the routing of a lightpath (both working and backup path if applicable), and the probability of every dual-span failure (based on the equipment and fiber making up those spans), one can compute the availability of a given service or lightpath. The recoverability metric $R(a, b)$ is dependent on the nature of the protection/restoration architecture used for the network and the lightpaths, as well as the sequence of events. In the next section, we develop a general Markov model that takes into account the sequence of events leading to service unavailability, as well as repair events.

12.3.2 A Markov Model Approach to Service Availability

The unavailability of all lightpaths is not simple to determine because of complex topologies, demand matrices, and actual routing of lightpaths and sharing of backup capacity. Furthermore, all connections do not have the same unavailability – longer lightpaths tend to have higher unavailability. We measure unavailability along the longest end-to-end lightpath.

To measure unavailability, we use a Markov model based on the sequences of transitions from working state to outage, or service unavailability, as shown in Figure 12.2 for DBPP and SBPP. For the dedicated backup case in Figure 12.2, an outage occurs, for example, as a result of a failure on the primary path (or backup path) followed by a failure on the backup path (or primary path) before the failed components on either path are repaired. For the shared backup case in Figure 12.2, an outage can occur as a result of a larger set of causes. In addition to the failure mode just described for dedicated backup path protection, a backup path can become unavailable because it is being used by some other lightpath whose primary path has failed [95, 169]. This cross-impact can be minimized by limiting sharing on shared backup channels [249] (see also Section 8.3). In fact, it is easy to realize that by limiting sharing to one primary path per shared backup channel, the availability of an SBPP lightpath becomes that of a DBPP lightpath. Other components of the model are channel protection and reprovisioning. Channel protection consists of recovering a failed primary path, in the case of a card or port failure, on an available channel on the same span or link where the failure occurred [16]. Consequently, the traffic is not switched on to the protection path. Reprovisioning allows a new backup path to be computed and provisioned to recover service immediately, in case switching to the original designated backup path fails.

The model shown in Figure 12.2 for a protected lightpath has five states where the service is available, A, A_1, A_2, B, and C. In the outage state D, the service is unavailable. There are two types of state transitions: those that correspond to failure events, and those that correspond to repair events. Among the transitions that correspond to a failure, some are accompanied by a recovery event such

---

[1]In practice, dual span failures are the most common failure modes that would contribute the most to service or lightpath unavailability.

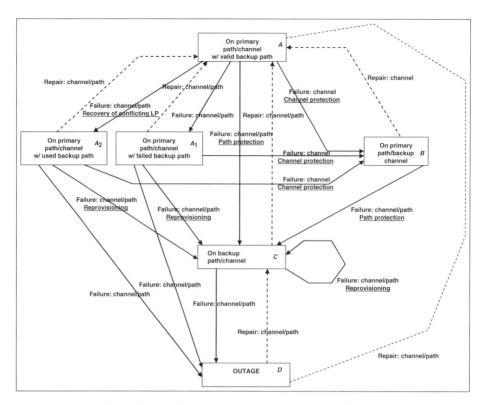

Figure 12.2: Path-protected mesh network: state behavior.

as channel protection, path protection, or reprovisioning, if the service would otherwise be affected. Some failure transitions are not accompanied by a recovery event if the failure does not cause a service impact, for example a failure of a channel on the backup path. Other failures are not accompanied by a recovery event because none of the recovery actions can restore service, typically after multiple failures without the corresponding repairs, leading to a service outage (state D). In state A, there is no failure that directly or indirectly invalidates any of the recovery mechanisms, and the service is on the primary path of the lightpath. In states A_1 and A_2, the service is still on the primary path but the backup path is unavailable because it has failed (A_2) or because it is being used by another lightpath whose primary has failed (A_1). Therefore, the backup path protection of the lightpath is unavailable. In state B, a channel failure has resulted in channel protection on the primary path as the means to recover from the failure. A subsequent and directly impacting failure on the primary path in states A and B causes a transition to state C after protection switching to the backup path. Any additional failure on the active path in states A_1, A_2, and C triggers a reprovisioning event, if this recovery scheme is available and succeeds. Otherwise, the lightpath becomes out-of-service in outage state D.

Figure 12.3 illustrates the different components making up a link between two switches that are used in the availability model. The number of components such as WDM mux/demux units and amplifiers depends on the link length. Multiple WDM systems can be concatenated using transponders on a link. The unavailability of an entire link, U_{link}, is derived from the unavailability of the key components responsible for a link failure, namely DWDM equipment, optical amplifiers and fiber. The unavailability of a single channel within a DWDM link, U_{channel}, is derived from the unavailability of the laser and receiver on that channel. The unavailabilities U_{link} and U_{channel} are calculated by

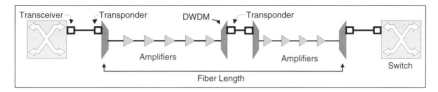

Figure 12.3: Components traversed by lightpath in availability model. (After [17], Figure 7. Reproduced by permission of © 2002 The Optical Society of America.)

adding the unavailability of all the serial components on the link or the channel, respectively (shown in Figure 12.3) as per Equation 12.9. Assuming that failure events in different links are independent, the unavailability of a path can be simply approximated as the sum of the unavailability of each link, or channel (if channel protection does not apply), making up that path.

Using the Markov model for path-protected mesh networks shown in Figure 12.2, one can compute the probability that a lightpath is unavailable. A lightpath becomes unavailable when the different recovery mechanisms in place, such as channel protection, dedicated or shared backup path protection, and reprovisioning, are unsuccessful in maintaining service, following the occurrence of multiple failures. Those failures occur according to the FIT rates or MTTF of the individual components, and based on the parallel and/or serial nature of their composition, and appear in the Markov model as failure rates λ. The unavailability is then obtained by observing that when a lightpath becomes unavailable, it remains in that state until some failed component that recovers service is repaired, which takes place according to the repair time MTTR of that component, and appears in the Markov model as repair rate μ.

12.3.3 Modeling Sharing of Backup Channels

Sharing of backup capacity or backup channels for shared backup path protection increases the chance that the shared backup capacity or channels will not be available for recovery in case of a failure of the primary path, as compared to dedicated backup path protection. This happens with a transition from state A to state A_1 when the backup path becomes unavailable because it is used by another lightpath whose primary path has failed. Removing state A_1 and the corresponding state transitions produces a Markov model for a dedicated backup protected lightpath. Some authors have modeled and analyzed the impact of sharing by assuming that all the primary paths sharing backup capacity or channel can fail independently. When this happens, they utilize the backup capacity or channel, thus preventing a successful recovery of any other lightpath sharing the backup capacity/channel in case of a second failure [95, 169].

12.3.4 Impact of Channel Protection

Channel protection consists of recovering a failed primary path in the case of a card or port failure on an available channel on the same span or link as discussed in detail in Chapter 3. Consequently, the traffic is not switched onto the protection path. We assume that channel protection occurs only once, after which path protection would be triggered. We also assume that only free channels (and not the available shared backup channels) are used for channel protection. Channel protection is modeled by the inclusion of state B and the corresponding transitions in the Markov model of Figure 12.2. The following parameters can be used to model the channel protection scheme. We let S be a decision variable, and we set it to 1 if channel protection is available, and to 0 otherwise. We also define s as the average ratio of free channels to working and protection (including shared backup) channels. Clearly, procuring additional lit capacity in the network increases s and the chance that channel protection will improve service availability in the network. Parameters S and s can be introduced in the Markov model of failure recovery in Figure 12.2 by setting the derived probabilities on the appropriate state transitions.

12.3.5 Impact of Reprovisioning

In the case of multiple failures, a mesh network utilizing intelligent OXCs can also support lightpath reprovisioning. Lightpath reprovisioning tries to establish a new backup path when protection on the original backup path does not succeed. Reprovisioning uses existing spare capacity and currently unused shared backup capacity to find a new backup path on which to immediately recover the failed lightpath. There are three conditions that result in lightpath reprovisioning: (a) a failure of the primary path followed by a failure of the backup path prior to repair of the primary path, (b) a failure of the backup path followed by a failure of the primary path prior to repair of the backup path, and (c) a failure of the primary path of a lightpath LP_1 sharing backup capacity with a lightpath LP_2, followed by a failure of the primary path of lightpath LP_2. This last case would cause a contention situation where more than one lightpath needs to use the shared backup capacity. In this case, lightpath LP_1 is recovered onto its backup path after the failure, thus occupying the shared backup resources. When lightpath LP_2 fails, it cannot recover onto its backup (because resources are being used), resulting in a reprovisioning attempt, although the likelihood of such a scenario occurring can be mitigated using real-time channel assignment strategies [89]. Note that reprovisioning may fail if there is not enough capacity available. However, the presence of lightpath reprovisioning increases the service availability of a mesh network. Service unavailability occurs as a result of multiple concurrent failure scenarios and the time it takes to fix the failure (e.g. hours if a fiber cut in a remote area needs to be repaired). Reprovisioning a lightpath that becomes unavailable after a double failure will improve the service availability of the network, by reducing the time that the service is unavailable from hours to tens of seconds. This is particularly significant in transoceanic networks where the time to repair a damaged undersea cable could be as much as 48 hours. Simulation studies have shown that, compared to traditional protection schemes, backup path protection can provide higher availability when complemented by reprovisioning after a second failure [17]. Note that reprovisioning can also be used in a preemptive manner by reprovisioning lightpaths following a failure in case they became unprotected and would not be able to recover from a subsequent failure [249].

Reprovisioning is modeled by the inclusion of state C and the corresponding transitions in the state transition diagram. The following parameters can be used to model reprovisioning. We let R be a decision variable, we set it to 1 if reprovisioning is available, and to 0 otherwise. We also define r as the average ratio of available capacity to working and protection (dedicated or shared backup) capacity. Clearly, procuring additional lit capacity in the network increases r and the chance that reprovisioning will succeed, therefore improving service availability in the network. Parameters R and r can be introduced in the Markov model of failure recovery in Figure 12.2 by setting the derived probabilities on the appropriate state transitions.

12.4 Service Availability in Path-Protected Single and Multi-Domain Mesh Networks

In this section inspired from the work in [17], we discuss and compare the path-protected mesh network availability model for single recovery domain and multiple recovery domain networks. For a single-domain network, the probability of failure along end-to-end primary and backup paths can be high, increasing the probability of service outage due to a double failure. A multi-domain network consists of independent routing and recovery domains. Routing and recovery are strictly limited to each domain with shorter primary and backup paths. This not only decreases the probability of failure along the primary and backup paths, but also decreases the probability that multiple concurrent failures cause a service outage, because these failures are likely to be spread across different recovery domains. We show that splitting a network into multiple domains increases the overall availability.

12.4.1 Network Recovery Architecture–Single Domain

Let us first consider the case of a single recovery domain. We assume that local channel protection is attempted prior to backup path protection. Figure 12.4 shows a reference primary path, LP_1, $\{A-B-C-D\}$, in a single-domain mesh network consisting of optical switches interconnected by WDM systems. The backup path for LP_1 is LB_1, $\{A-E-F-G-H-I-D\}$. We consider the outage scenario for LP_1. There are essentially two types of failures on LP_1. The first affects only a single channel. An example is the failure of a laser or a receiver on any of the interface ports either on the optical switch or the transponder in the WDM system on the lightpath. The other type of failure is that of multiple channels failing due to a common component failure such as a fiber cut or an optical amplifier (OA) failure.

We make the following assumptions regarding the different recovery mechanisms. First, lightpath recovery is attempted through channel protection, but only once, using free channels on that link, not shared backup channels. Second, path protection is attempted. Third, reprovisioning is applied as many times as possible. When the first type of failure occurs, the lightpath is recovered through channel protection, using another optical channel within the same link. However, for the second type of failure the lightpath can only be recovered from the end-nodes, using backup path protection. For example, if any one of the individual lasers or receivers on the channel in link $A-B$ used by LP_1 fails, the lightpath may recover locally by taking up a channel on the same link. On the other hand, if the entire link $A-B$ fails, the lightpath is recovered on LB_1. If LB_1 is also unavailable, a new backup path is computed by the Network Management System (NMS), or by the end-nodes, based on the available capacity in the network, so that the traffic may be recovered. If no such alternate route with spare capacity is found, the recovery process ends and the service becomes unavailable. The Markov model of failure events (such as laser or optical amplifier failures) and recovery events (such as channel protection, path protection, reprovisioning) is shown in Figure 12.2.

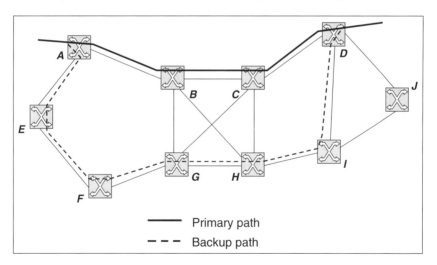

Figure 12.4: Single-domain network. (From [17], Figure 3. Reproduced by permission of © 2002 The Optical Society of America.)

12.4.2 Network Recovery Architecture–Multiple Domains

Let us now consider the case of a multi-domain recovery architecture. For a single-domain network, the backup path can be routed through any portion of the entire network. Thus, it is possible that a local failure triggers a very long recovery event since the backup path itself may traverse the entire length of the network (see Figure 12.4). Furthermore, the probability of failure along a longer path is higher since a longer path would traverse more components prone to failure (fiber, WDM systems, amplifiers, transceivers, etc.).

A multi-domain network consists of independent routing and recovery domains. The domains are typically connected by two gateway nodes, a primary and a secondary, as shown in Figure 12.5. The primary gateway acts as the recovery point for cable, equipment and intermediate node failures, whereas the secondary gateway node is used to recover in case of the failure of the primary gateway node (which is handled via lightpath reprovisioning). Thus, the end-to-end lightpath consists of smaller lightpath segments that are routed to/from and between the gateway nodes. Routing and recovery are strictly limited to each domain, thus the backup paths for local failures will be contained in the appropriate domain and will not traverse into neighboring domains [17]. This approach results in shorter local lightpaths and faster recovery. The resulting end-to-end primary and backup paths may be longer than the shortest paths in the single-domain network; and the multi-domain mesh network may achieve less sharing and thus require more capacity than the single-domain mesh network. However, the probability of failure is smaller along the shorter lightpath segments that are confined to a single domain. Furthermore, it is less likely that multiple failures will occur in the same domain, further reducing the chances of end-to-end service outage.

For the multi-domain case, the unavailability is first computed for each domain since recovery is restricted to each domain. The overall unavailability is then calculated as the sum of the unavailabilities for each domain.

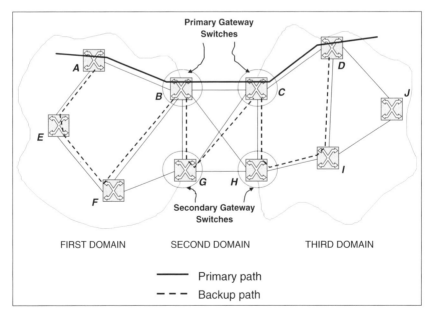

Figure 12.5: Multi-domain network. (From [17], Figure 4. Reproduced by permission of © 2002 The Optical Society of America.)

12.4.3 Results and Discussion

We consider a transoceanic network that spans the United States, the Atlantic ocean and Europe. In the single-domain approach, this network is viewed as a single flat network. In the multi-domain approach, the network consists of the North American, Atlantic and European domains. As in Figure 12.6, there are two gateways at each domain border. In our experimental network, there are primary gateways at New York and London; and secondary gateways at Philadelphia and Paris. This network is a hypothetical transatlantic carrier's network.

As discussed above, we use the longest end-to-end path to measure the unavailability of the network. Furthermore, we assume that the reprovisioning success probability is 0.5. The FIT, MTTR and other parameters for the components on the links are given in Tables 12.1 and 12.2. Note that we assume that the transoceanic equipment (optical amplifiers, fibers) are 10 times more reliable than the corresponding terrestrial equipment.

We consider an end-to-end lightpath from San Francisco to Copenhagen as shown in Figure 12.6 [17]. In the single-domain mode, both the primary and backup paths traverse all three domains without any constraints. In the multi-domain mode, the lightpath consists of segments from San Francisco to

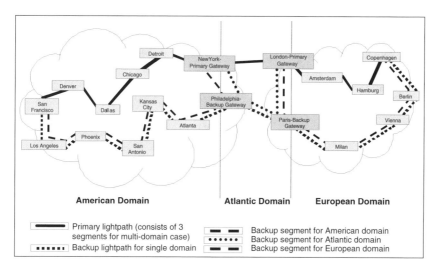

Figure 12.6: Primary and backup paths in hypothetical network for single and multi-domains. (From [17], Figure 9. Reproduced by permission of © 2002 The Optical Society of America.)

Table 12.1: Component FIT rates. (After [17], Table 1. Reproduced by permission of © 2002 The Optical Society of America.)

Components	FIT
Terrestrial optical amplifier	2000
Transoceanic optical amplifier	200
Mux/demux unit	2000
Transponder	1500
Terrestrial fiber cut/km	50
Transoceanic fiber cut/km	5

Table 12.2: Component MTTR and other parameters. (After [17], Table 2. Reproduced by permission of © 2002 The Optical Society of America.)

Terrestrial optical amplifier spacing	80 km
Transoceanic optical amplifier spacing	50 km
# of spans/terrestrial WDM	7
# of WDM per link	1
Terrestrial fiber MTTR	4 hrs
Transoceanic fiber MTTR	48 hrs
Terrestrial equipment MTTR	4 hrs
Transoceanic equipment MTTR	48 hrs
Reprovisioning success probability	0.5

Table 12.3: Hop and path lengths for primary and backup path segments. (After [17], Table 3. Reproduced by permission of © 2002 The Optical Society of America.)

	Primary hops	Primary length (km)	Backup hops	Backup length (km)
Single-Domain Net	9	19100	11	21300
Multi-Domain Net				
North American	5	7450	7	8250
Trans-Atlantic	1	10250	3	11250
European	3	1650	5	3800
Overall	9	19350	15	23300

New York, New York to London, and London to Copenhagen. For each of these lightpath segments, recovery is performed in the corresponding domain. Table 12.3 lists the hop length and path length of the primary and backup paths for both the single-domain and multi-domain cases.

The availability calculations were performed in [17] using a Markov model similar to the one introduced in Section 12.3 and depicted in Figure 12.2 for dedicated backup and shared backup path protection. For shared backup path protection, reprovisioning was complementing path protection, and the availability was computed with and without channel protection enabled. For dedicated backup path protection, reprovisioning and channel protection were not considered. The unavailability and corresponding availability results are given in Table 12.4.

The results in Table 12.4 illustrate four factors that affect network availability. First, the availability depends heavily on the length of the lightpath and the time to repair a failure (MTTR). The long trans-Atlantic lightpath segment, which has higher equipment MTTRs, clearly dominates in terms of unavailability. Second, availability depends on the recovery mechanisms. Compared to dedicated backup path protection, reprovisioning after a second failure in the case of shared backup path protection,[2] provides higher availability, resulting in up to a 48% decrease in unavailability [17]. Third, a further increase in availability is achieved if channel protection is implemented prior to backup path protection. Fourth and last, implementing a multi-domain network rather than a single-domain network increases overall availability as failures in different domains are recovered independently, resulting in up to a 55% decrease in unavailability [17]. In summary, using shared backup path protection in conjunction with channel protection and reprovisioning in a multi-domain network resulted

---

[2]Clearly, reprovisioning could also be invoked in the case of dedicated backup path protection, but that was not the case in the work described here [17].

Table 12.4: Unavailability and availability results. (After [17], Table 4. Reproduced by permission of © 2002 The Optical Society of America.)

	Dedicated backup path protection		Shared backup path protection w/ reprovisioning – no channel protection		Shared backup path protection w/ reprovisioning – channel protection	
	Unavail-ability. (min/yr.)	Avail-ability. (%)	Unavail-ability. (min/yr.)	Avail-ability. (%)	Unavail-ability. (min/yr.)	Avail-ability. (%)
Single-Domain Net	33.7	99.994	18.3	99.997	17.5	99.997
Multi-Domain Net						
North American	3.12	99.9994	1.82	99.9997	1.68	99.9997
Transatlantic	12.0	99.998	6.19	99.999	6.15	99.999
European	0.41	99.99992	0.25	99.99995	0.22	99.99996
Overall	15.53	99.997	8.26	99.998	8.05	99.998

in a 76% combined decrease in unavailability compared to dedicated backup path protection in the single-domain case [17].

12.4.4 A Simple Model

To gain a basic understanding of the availability of protected circuits, we propose the following simple model, where the unavailability of a path is proportional to its length, and a lightpath becomes unavailable when both the primary and backup paths are unavailable. Let l_1 be the length of the primary path, and let $l_2 \geq l_1$ be the length of the backup path. Then the unavailability of an SBPP or DBPP lightpath in a single domain can be approximated as

$$U_{\text{mesh-sd}} \simeq l_1 \times l_2 \geq l_1^2 \qquad (12.10)$$

Using this simple model for the case of multiple domains, we let l_{1k} and l_{2k} be the lengths of the primary and backup paths respectively, for the kth domain. The lightpath will be unavailable if any of the lightpath segments are unavailable, yielding the approximation

$$U_{\text{mesh-md}} \simeq \sum_k l_{1k} \times l_{2k} \geq \sum_k l_{1k}^2 \qquad (12.11)$$

Now, assuming that $l_1 \approx \sum_k l_{1k}$ and $l_2 \approx \sum_k l_{2k}$, it follows, based on the simple algebraic argument $(\sum_k l_{1k})(\sum_k l_{2k}) \geq \sum_k l_{1k} \times l_{2k}$, that the unavailability in single-domain mesh networks is higher than the unavailability in multi-domain mesh networks, $U_{\text{mesh-sd}} \geq U_{\text{mesh-md}}$, everything else being equal.

12.5 Service Availability in Ring-Protected and Path-Protected Networks

The focus of this section has been an unexplored dimension in the mesh versus ring debate, until recently. There has been a perception in the industry that ring-protected networks were more reliable than mesh-protected networks. This perception was based on the assumption that fast recovery from a failure leads to shorter downtime and hence higher availability. Using the work in [275], we show that this is not necessarily the case, and explain why.

12.5.1   Ring Availability Analysis

In the ring architecture, we assume that the network is divided into a set of suitable rings. For the availability model, we need to consider how a lightpath is routed through such a ring-based network. We assume that the average ring circumference is 2000 km. A BLSR ring of up to 2000 km guarantees 60 ms recovery time. In an N-node ring, the shortest path between two nodes has at most $N/2$ hops and at least 1 hop. Hence, the average hop count of a working path in a ring is $N_{\text{hops}} = N/4$. With a ring circumference of size 2000 km, the average number of rings traversed by a lightpath of length L (in km) is $N_r = L/500$. The failures in these rings are independent, and the end-to-end lightpath fails when any of the rings it traverses fails. Hence, we can concentrate on a single ring and then multiply the unavailability for one ring by the number N_r of rings traversed. Adjacent rings are assumed to be connected by SONET/SDH APS 1+1 protection switching configuration in a single place (Figure 12.7), and therefore dual-ring interconnect is not assumed. A lightpath traverses $(N_r - 1)$ of these ring interconnects. The failure of any one of these configurations also leads to the failure of the lightpath. Thus, the unavailability of the lightpath is given by [275]:

$$N_r \times (\text{ring unavailability}) + (N_r - 1) \times (\text{APS 1+1 unavailability}) \qquad (12.12)$$

Figure 12.7 illustrates the failure scenarios in rings. A protection link protects a primary link on each span around the ring. When the primary link on a span fails, the protection link is used. When the protection link fails, or when both the working and protection links on a span fail due to fiber cut or amplifier/DWDM failure, the other (longer) segment of the ring (connecting the endpoints of this span) is used to recover this span in a hairpin-like arrangement. Consider a path on a ring traversing consecutive nodes A, B, C, and D in that order. Suppose span $B-C$ fails. Then, the recovered traffic traverses the following path: span $A-B$, the larger circumference of the ring from B to C, and finally span $C-D$. Thus, the recovery path has a hop count of $(N_{\text{hops}} - 1) + (N - 1) = (N_{\text{hops}} + N - 2)$. Note that for the portion of the recovery path – the hairpin portion – which overlaps with the original path (in this case spans $A-B$ and $C-D$), traffic flows on the working link in one direction and on the protection link in the other direction. The remaining portion of the recovered path (in this case, the larger circumference segment from B to C) uses the protection link on each span, the working link of which carries other traffic. Thus, the failure of any link (primary/working or protection) on the protection path leads to loss of service for the lightpath.

A Markov model for a lightpath in a ring to transition from the working state to unavailable state can be found in [275].

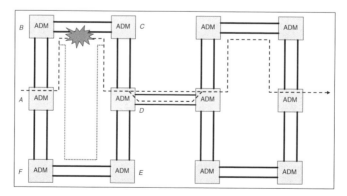

Figure 12.7: Failure scenarios in rings. (After [275], Figure 4. Reproduced by permission of © 2001 The Institute of Electronics, Information and Communication Engineers.)

Note also that in a ring environment, without the flexibility of routing and automatic topology discovery that is inherent in the new generation of optical mesh networks being deployed [38], it is not feasible to reprovision after a lightpath becomes unavailable.

12.5.2   Results and Discussion

The most important result we report from [275] is that the unavailability of an end-to-end circuit increases roughly linearly in a network architecture that is composed of multiple rings, whereas it increases roughly as the square of the path length in a single-domain mesh network. This is due to the fact that in single-domain mesh networks, unavailability is proportional to the product of both the primary and backup path lengths and approximately to the square of the path length. A ring network on the other hand, consists of one or more interconnected rings. Consequently, the unavailability of a circuit in a ring network is proportional to the number of rings it goes through, which itself is proportional to the length of the circuit [275]. In Figure 12.8, unavailability in minutes/year is plotted as a function of lightpath length. The same data are plotted as the percentage of availability as a function of lightpath length in Figure 12.9.

For shorter path lengths, the unavailability in a ring network is higher than the unavailability in a mesh network, because recovery around the ring must typically traverse a longer distance than in a mesh network. However, the availability in a single-domain mesh network will decrease faster than that in a network consisting of multiple rings as distances increase, and will cross over at some path length. The authors in [275] show that the cross-over takes place around 2500 km, at which point a single-domain mesh network becomes less reliable than a ring network. They also show that for lightpaths of length up to 4200 km, both single-domain mesh and ring networks can meet 99.999% availability requirements. In addition, if a mesh network is divided into suitable multiple domains where failure recovery is confined within a domain, then a multi-domain mesh network can always be more reliable than an interconnected multi-ring network, regardless of the end-to-end circuit length.

We summarize the important observations from the work in [275] as follows:

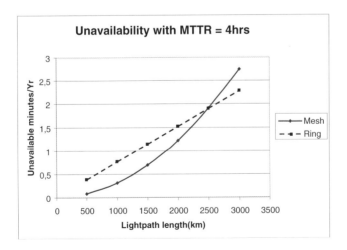

Figure 12.8: Unavailability in single-domain mesh networks and ring networks. (From [275], Figure 6. Reproduced by permission of © 2001 The Institute of Electronics, Information and Communication Engineers.)

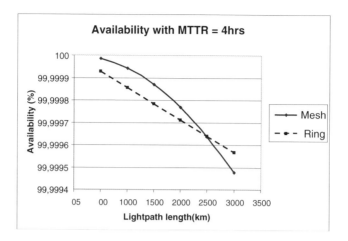

Figure 12.9: Availability in single-domain mesh networks and ring networks. (From [275], Figure 7. Reproduced by permission of © 2001 The Institute of Electronics, Information and Communication Engineers.)

- The unavailability of a lightpath in a ring network is proportional to its length. In a single-domain mesh network, the unavailability varies as the product of the lengths of the primary and backup paths, which can be approximated as the square of the length of the lightpath.

- For lightpaths of length up to 4200 km, both single-domain mesh networks and ring networks were shown to meet 99.999% availability requirements.

- For networks spanning up to 2500 km, a single-domain mesh network is more reliable than a ring network. For networks spanning more than 2500 km, a mesh network can be made more reliable than a ring network if it is divided into multiple domains, each of which spanning 2500 km or less.

12.5.3 The Simple Model Again

Again, using the simple model we introduced earlier in the chapter, we let l_{1k} and $l_{3k} \geq l_{2k}$ be the length of the primary paths[3] and backup paths respectively, for the kth ring. The lightpath will be unavailable if any of the lightpath segments are unavailable, yielding the approximation:

$$U_{\text{ring}} \simeq \sum_k l_{1k} \times l_{3k} \tag{12.13}$$

Now, assuming that $l_1 \approx \sum_k l_{1k}$ and with $l_3 = \sum_k l_{3k}$, it follows that (1) the unavailability in single-domain mesh networks is lower than the unavailability in single-ring networks (small geographical span) because $l_3 \geq l_2$, and (2), based on simple algebraic arguments, the unavailability in single-domain mesh networks is higher than the unavailability in multi-ring networks (large geographical span), everything else being equal.

Recovery time and network availability are clearly two mostly independent issues. Thus, even if mesh networks have higher recovery latencies, they can be designed to be more reliable than ring networks. This dispels the common myth that rings are more reliable due to lower recovery times.

---

[3]We assume that the length of the primary path in one of the rings is roughly the same as the length of the primary path in one of the recovery domains in a multi-domain mesh network.

12.6 Conclusion

In this chapter, we introduced the notion of network service availability, focusing on dual-failure scenarios. We presented different recovery mechanisms in single-domain mesh networks (e.g. dedicated and shared backup path protection, channel protection, reprovisioning) and discussed how they impact on the overall service availability performance. We introduced a Markov model that takes into account these different recovery mechanisms. We then presented results comparing single and multi-domain mesh networks, in the context of a trans-oceanic network spanning multiple continents.

Despite the widespread belief that networks with faster recovery times achieve higher service availability, availability is, instead, heavily influenced by the recovery mechanisms. Furthermore, availability may have little to do with recovery speed when the recovery time is small compared to the time needed to replace the failed elements or to repair a fiber cut. We presented results comparing the availability in mesh and ring-based networks, and concluded that for limited geographical spans, single-domain mesh networks were more reliable than ring-based networks, but that for larger geographical spans, single-domain mesh networks were less reliable than networks of interconnected rings, but could be made more reliable by dividing them into multiple recovery domains.

Bibliography

[1] *Automatic Protection Switching for SONET*. Technical Report SR-NWT-001756, Bell Communications Research, Red Bank, NJ, October 1990.

[2] *Generic Reliability Assurance Requirements for Fiber Optic Transport Systems*. Technical Report TR-NWT-000418, Bell Communications Research, Red Bank, NJ, 1990.

[3] *Reliability and Quality Switching Systems Generic Requirements*. Technical Report TA-NWT-000284, Bell Communications Research, Red Bank, NJ, 1991.

[4] *Synchronous Optical Network (SONET) Transport Systems: Common Generic Criteria*. Technical Report TR-NWT-000253, Bell Communications Research, Red Bank, NJ, December 1991.

[5] *Reliability Assurance Practices for Optoelectronic Devices in Interoffice Applications*. Technical Report TR-NWT-000468 Issue 1, Bell Communications Research, Red Bank, NJ, 1992.

[6] *An Introduction to the Reliability and Quality Generic Requirements*. Technical Report TR-NWT-000874 Issue 4, Bell Communications Research, Red Bank, NJ, 1993.

[7] *Generic Requirements for Assuring the Reliability of Components Used in Telecommunications Equipment*. Technical Report TR-NWT-000357 Issue 2, Bell Communications Research, Red Bank, NJ, 1993.

[8] *Transport Systems: Generic Requirements*. Technical Report TR-NWT-000499, Bell Communications Research, Red Bank, NJ, December 1993.

[9] *SONET Dual-Fed Unidirectional Path Switched Ring (UPSR) Equipment Generic Criteria*. Technical Report GR-1400-CORE, Bell Communications Research, Red Bank, NJ, March 1994.

[10] *SONET Bidirectional Line-Switched Ring Equipment Generic Criteria*. Technical Report GR-1230-CORE, Bell Communications Research, Red Bank, NJ, December 1996.

[11] *Planning for Service Survivability in Broadband Multilayer Networks*. Technical Report SR-4317, Bell Communications Research, Red Bank, NJ, 1997.

[12] *User Network Interface (UNI) v1.0 Signaling Specification*. Technical Report OIF Implementation Agreement OIF-UNI-01.0, October 2001.

[13] G. P. Agrawal. *Fiber-Optic Communication Systems*. John Wiley & Sons, New York, 1997.

Path Routing in Mesh Optical Networks   Eric Bouillet, Georgios Ellinas,
Jean-François Labourdette, Ramu Ramamurthy   © 2007 John Wiley & Sons, Ltd

[14] R. K. Ahuja, T. L. Magnanti, and J. B. Orlin. *Network Flows*. Prentice-Hall, Englewood Cliffs, NJ, 1993.

[15] M. Aigner. *Graph Theory – A Development from the 4-color Problem*. BCS Associates, Moscow, Idaho, 1986.

[16] A. A. Akyamaç et al. Optical mesh networks modeling: Simulation and analysis of restoration performance. In *Proceedings of the 18th Annual National Fiber Optic Engineers Conference (NFOEC)*, Dallas, TX, September 2002.

[17] A. Akyamaç et al. Reliability of single domain versus multi domain optical mesh networks. In *Proceedings of the 18th Annual National Fiber Optic Engineers Conference (NFOEC)*, Dallas, TX, September 2002.

[18] A. Akyamaç et al. Ring speed restoration and optical core mesh networks. In *Proceedings of the 7th European Conference on Networks and Optical Communications (NOC)*, Darmstadt, Germany, June 2002.

[19] W. T. Anderson. The MONET project final report. In *Proc. IEEE/OSA Optical Fiber Commun. Conf. (OFC)*, pages 148–149, Baltimore, MD, February 2000.

[20] D. W. Anding. Service offerings utilizing digital cross-connect systems – A Bell South perspective. In *Proc. IEEE Int. Conf. Commun. (ICC)*, pages 336–339, Philadelphia, PA, June 1998.

[21] N. Antoniades, I. Roudas, G. Ellinas, and J. Amin. Transport metropolitan optical networking: Evolving trends in the architecture design and computer modeling. *IEEE/OSA J. Lightwave Technology*, 22(11): 2653–2670, November 2004.

[22] N. Antoniades, M. Yadlowsky, and V. L. daSilva. Computer simulation of a metro WDM ring network. *IEEE Photonics Technology Letters*, 12(11): 1576–1578, 2000.

[23] A. Antonopoulos et al. Design methodology for ETSI SDH subnetworks employing SNCP-rings. In *Proc. IEEE Int. Conf. Commun. (ICC)*, pages 878–882, Montreal, Canada, June 1997.

[24] G. R. Ash. *Dynamic Routing in Telecommunications Networks*. McGraw-Hill, New York, 1997.

[25] G. R. Ash, Kenneth Chan, and J.-F. P. Labourdette. Analysis and design of fully shared networks. In *14th International Teletraffic Congress*, Antibes Juan-Les-Pins, France, June 1994.

[26] L. Berger et al. *Generalized MPLS – Signaling Functional Description*. Internet Engineering Task Force (IETF), Technical Report RFC 3471, January 2003.

[27] L. Berger et al. *Generalized MPLS RSVP-TE Extensions*. Internet Engineering Task Force (IETF), Technical Report RFC 3473, January 2003.

[28] J. Augusto, S. Monteiro, and M. Gerla. Topological reconfiguration of ATM networks. In *Proc. IEEE Infocom.*, San Francisco, CA, June 1990.

[29] G. Austin et al. Fast, scalable, and distributed restoration in general mesh optical networks. *Bell Labs Technical Journal*, 6(1): 67–81, 2001.

[30] K. Bala et al. The case for opaque multiwavelength lightwave networks. In *IEEE/LEOS Summer Topical Meeting*, Keystone, Colorado, August 1995.

[31] I. Baldine and G. N. Rouskas. Dynamic reconfiguration policies for WDM networks. In *Proc. IEEE Infocom.*, New York, NY, March 1999.

[32] D. Banerjee and B. Mukherjee. Wavelength-routed optical networks: Linear formulation, resource budgeting tradeoffs, and a reconfiguration study. In *Proc. IEEE Infocom.*, Kobe, Japan, April 1997.

[33] R. Batchellor. *Coordinating Protection in Multiple Layers*.Technical Report oif99.038.0, Optical Inter-networking Forum (OIF), April 1999.

[34] R. Bellman. On a routing problem. *Quarterly of Applied Mathematics*, 16(1): 87–90, 1958.

[35] V. Benes. Optimal rearrangeable multistage connecting networks. *The Bell System Technical Journal*, 43(7): 1641–1656, 1964.

[36] D. Benjamin, R. Trudel, S. Shew, and E. Kus. Optical services over an intelligent optical network. *IEEE Communications Mag.*, pages 73–78, September 2001.

[37] J. C. Bermond, B. Jackson, and F. Jaeger. Shortest coverings of graphs with cycles. *J. Combinatorial Theory*, Ser. B 35: 297–308, 1983.

[38] G. Bernstein, B. Rajagopalan, and D. Saha. *Optical Network Control: Architecture, Protocols and Standards*. Addison-Wesley, Reading, Mass., 2004.

[39] G. Bernstein, J. Yates, and D. Saha. IP-centric control and management of optical transport networks. *IEEE Communications Mag.*, pages 161–167, October 2000.

[40] R. Bhandari. *Survivable Networks: Algorithms for Diverse Routing*. Kluwer Academic Publishers, 1999.

[41] M. Bhardwaj, L. McCaughan, S. K. Korotky, and I. Saniee. Global expectation values of shared restoration capacity for general mesh networks. In *Proc. IEEE/OSA Optical Fiber Commun. Conf. (OFC)*, Los Angeles, CA, March 2004.

[42] M. Bhardwaj, L. McCaughan, S. K. Korotky, and I. Saniee. Analytical description of shared restoration capacity for mesh networks. *J. Optical Network*, 4(3): 130–141, 2005.

[43] M. Bhardwaj, L. McCaughan, A. Olkhovets, and S. K. Korotky. Simulation and modeling of the mean time to restore for optical mesh networks. In *Proc. IEEE/OSA Optical Fiber Commun. Conf. (OFC)*, Anaheim, CA, March 2005.

[44] R. Bhatia, M. Kodialam, and T.V. Lakshman. Fast network reoptimization schemes for MPLS and optical networks. *Computer Networks*, 50(3): 317–331, 2006.

[45] J. Bicknell, C. E. Chow, and S. Syed. Performance analysis of fast distributed network restoration algorithms. In *Proc. IEEE Globecom.*, pages 1596–1600, Houston, TX, November 1993.

[46] S. Biswas, S. Chaudhuri, S. Datta, K. Hua, and S. Sengupta. StarNet Modeler: A hybrid modeling system for path restoration in optical networks. In *OPNETWORK*, Washington, DC, August 2001.

[47] M. Boduch *et al.* Transmission of 40 Gbps signals through metropolitan networks engineered for 10 Gbps signals. In *Proc. IEEE/OSA Optical Fiber Commun. Conf. (OFC)*, Anaheim, CA, March 2006.

[48] Z. R. Bogdanowicz. Analysis of optimal sets of survivable paths in undirected simple graph applicable for optical networks. *Electronic Notes on Discrete Mathematics (ENDM)*, 22: 1–5, 2005.

[49] Z. R. Bogdanowicz and S. Datta. Analysis of backup route reoptimization algorithms for optical shared mesh networks. *Math. Comput. Modelling*, 40(11–12): 1047–1055, 2004.

[50] P. A. Bonenfant and C. M. C. Davenport. *SONET Optical Layer Protection Switching*. Technical Report TM-24609, Bell Communications Research, Red Bank, NJ, December 1994.

[51] E. Bouillet *et al*. Stochastic approaches to compute shared mesh restored lightpaths in optical network architectures. In *Proc. IEEE Infocom.*, New York, NY, June 2002.

[52] E. Bouillet and J. Labourdette. Distributed computation of shared backup path in mesh optical networks using probabilistic methods. *IEEE/ACM Trans. Networking*, 12(5): 920–930, 2004.

[53] E. Bouillet, J. Labourdette, and S. Biswas. Impact of multi-port card diversity constraints in mesh optical networks. In *Proceedings of the Annual National Fiber Optic Engineers Conference (NFOEC)*, Anaheim, CA, March 2006.

[54] E. Bouillet, J. Labourdette, R. Ramamurthy, and S. Chaudhury. Lightpath reoptimization in mesh optical networks. *IEEE/ACM Transactions on Networking*, 13(2): 437–447, April 2005.

[55] E. Bouillet, J-F. Labourdette, R. Ramamurthy, and S. Chaudhuri. Enhanced algorithm cost model to control tradeoffs in provisioning shared mesh restored lightpaths. In *Proc. IEEE/OSA Optical Fiber Commun. Conf. (OFC)*, Anaheim, CA, March 2002.

[56] E. Bouillet, J.F. Labourdette, G. Ellinas, R. Ramamurthy, and S. Chaudhuri. Local optimization of shared backup channels in optical mesh networks. In *Proc. IEEE/OSA Optical Fiber Commun. Conf. (OFC)*, Atlanta, GA, March 2003.

[57] E. Bouillet, P. Mishra, J.-F. Labourdette, K. Perlove, and S. French. Lightpath reoptimization in mesh optical networks. In *Proceedings of the 7th European Conference on Networks and Optical Communications (NOC)*, Darmstadt, Germany, June 2002.

[58] C. A. Brackett. Dense wavelength division multiplexing networks: Principles and applications. *IEEE J. Select. Areas Commun.*, 8(6): 948–964, August 1990.

[59] D. Brélaz. New methods to color the vertices of a graph. *Communications of the ACM*, 22(4): 251–256, April 1979.

[60] J. Carlier and D. Nace. A new distributed restoration algorithm for telecommunication networks. In *Proceedings of the Sixth International Network Planning Symposium*, Budapest, Hungary, September 1994.

[61] P. A. Catlin. Double cycle covers and the Petersen graph. *J. Graph Theory*, 13(4): 465–483, 1989.

[62] U. Celmins. On conjectures relating to snarks. PhD thesis, University of Waterloo, 1987.

[63] G. K. Chang, G. Ellinas, J. K. Gamelin, M. Z. Iqbal, and C. A. Brackett. Multiwavelength reconfigurable WDM/ATM/SONET network testbed. *IEEE/OSA J. Lightwave Technology*, 14(6): 1320–1340, 1996.

[64] C-W. Chao, P. M. Dollard, J. E. Weythman, L. T. Nguyen, and H. Eslambolchi. FASTAR – A robust system for fast DS3 restoration. In *Proc. IEEE Globecom.*, pages 1396–1400, Phoenix, AZ, December 1991.

[65] P. Charalambous, G. Ellinas, C. Dennis, E. Bouillet, J.F. Labourdette, S. Chaudhuri, M. Morokhovich, and D. Shales. Dynegy's national long-haul optical mesh network utilizing intelligent optical switches. In *Proc. LEOS Annual Meeting*, Glasgow, Scotland, November 2002.

[66] P. Charalambous *et al*. A national mesh network using optical cross-connect switches. In *Proc. IEEE/OSA Optical Fiber Commun. Conf. (OFC)*, Atlanta, GA, March 2003.

[67] G. Chartrand and L. Lesniak-Foster. *Graphs and Digraphs*. Chapman and Hall, New York, 1996.

[68] S. Chaudhuri, E. Bouillet, and G. Ellinas. Addressing transparency in DWDM mesh survivable networks. In *Proc. IEEE/OSA Optical Fiber Commun. Conf. (OFC)*, Anaheim, CA, March 2001.

[69] D.Z. Chen *et al*. World's first 40 Gbps overlay on a field-deployed, 10 Gbps, mixed-fiber, 1200 km, ultra long-haul system. In *Proc. IEEE/OSA Optical Fiber Commun. Conf. (OFC)*, Los Angeles, CA, March 2005.

[70] D.Z. Chen *et al*. New field trial distance record of 3040 km on wide reach WDM with 10 and 40 Gbps transmission including OC-768 traffic without regeneration. In *Proc. IEEE/OSA Optical Fiber Commun. Conf. (OFC)*, Anaheim, CA, March 2006.

[71] A. Chiu and J. Strand. Joint IP/Optical layer restoration after a router failure. In *Proc. IEEE/OSA Optical Fiber Commun. Conf. (OFC)*, Anaheim, CA, March 2001.

[72] A. Chiu and J. Strand. An agile optical layer restoration method for router failures. *IEEE Network Mag.*, pages 38–42, March/April 2003.

[73] C-H. Chow, S. Syed, J. Bicknell, and S. McCaughey. System and method for restoring a telecommunications network based on a two-prong approach, US Patent 5,495,471 February 1996.

[74] C-H. E. Chow, S. McCaughey, and S. Syed. *RREACT: A Distributed Protocol for Rapid Restoration of Active Communication Trunks*. Technical Report EAS-CS-92-18, UCCS, November 1992.

[75] E. Chow, S. McCaughey, and S. Syed. RREACT: A distributed network restoration protocol for rapid restoration of active communication trunks. In *Proc. Second IEEE Network Management and Control Workshop*, pages 391–406, Tarrytown, NY, September 1993.

[76] C. Clos. A study of non-blocking switching networks. *The Bell System Technical Journal*, 32(2): 406–424, 1953.

[77] M. Clouqueur and W. D. Grover. Availability analysis of span-restorable mesh networks. *IEEE J. Select. Areas Commun.*, 20(4): 810–821, 2002.

[78] M. Clouqueur and W. D. Grover. Mesh-restorable networks with complete dual failure restorability and with selectively enhanced dual-failure restorability properties. In *Proceedings of SPIE OptiComm*, pages 1–12, Boston, MA, July 2002.

[79] M. Clouqueur and W. D. Grover. Quantitative comparison of end-to-end availability of service paths in ring and mesh-restorable networks. In *Proceedings of the 19th National Fiber Optic Engineers Conference, NFOEC*, pages 317–326, Orlando, FL, September 2003.

[80] M. Clouqueur and W.D. Grover. Availability analysis and enhanced availability design in p-cycle-based networks. *Photonic Network Communications*, 10(1): 55–71, 2005.

[81] M. Clouqueur and W.D. Grover. Mesh-restorable networks with enhanced dual-failure restorability properties. *Photonic Network Communications*, 9(1): 7–18, 2005.

[82] B. Coan, W. E. Leland, M. P. Vecchi, A. Weinrib, and L. T. Wu. Using distributed topology update and preplanned configurations to achieve trunk network survivability. *IEEE Trans. Reliability*, 40(4): 404–416, 1991.

[83] D. Colle *et al*. Data-centric optical networks and their survivability. *IEEE J. Select. Areas Commun.*, 20(1): 6–20, 2002.

[84] R. Coltun. *The OSPF Opaque LSA Option*. Technical Report RFC 3472, Internet Engineering Task Force(IETF), July 1998.

[85] J. K. Conlisk. Topology and survivability of future transport networks. In *Proc. IEEE Globecom.*, pages 826–834, Dallas, TX, November 1989.

[86] T. H. Cormen, C. E. Leiserson, and R. L. Rivest. *Introduction to Algorithms*. McGraw-Hill, New York, 1990.

[87] B. Cortez. The emerging intelligent optical network: Now a reality. In *Proc. IEEE/OSA Optical Fiber Commun. Conf. (OFC)*, pages 224–225, Anaheim, CA, March 2002.

[88] S. Datta, S. Biswas, S. Sengupta, and D. Saha. Routing and grooming in two-tier survivable optical mesh networks. In *Proc. International Workshop on Quality of Service (IWQoS)*, Monterey, CA, June 2003.

[89] S. Datta, S. Sengupta, S. Biswas, and S. Datta. Efficient channel reservation for backup paths in optical mesh networks. In *Proc. IEEE Globecom.*, San Antonio, TX, November 2001.

[90] R. P. Davidson and N. J. Muller. *The Guide to SONET: Planning, Installing and Maintaining Broadband Networks*. Telecom Library, New York, 1983.

[91] P. Demeester *et al*. Resilience in multilayer networks. *IEEE Communications Mag.*, pages 70–76, August 1999.

[92] E. W. Dijkstra. A note on two problems in connexion with graphs. *Numerische Mathematik*, 1: 269–271, 1959.

[93] B. Doshi *et al*. Optical network design and restoration. *Bell Labs Technical Journal*, 4(1): 58–84, 1999.

[94] B. Doshi, R. Nagarajan, and M. Qureshi. Fast distributed restoration in general mesh optical networks. In *Proceedings of the 17th Annual National Fiber Optic Engineers Conference (NFOEC)*, Baltimore, MD, July 2001.

[95] B. T. Doshi, D. R. Jeske, N. Raman, and A. Sampath. Reliability and capacity efficiency of restoration strategies for telecommunication networks. In *Proc. Design of Reliable Communication Networks (DRCN)*, Banff, Canada, October 2003.

[96] J. Doucette, M. Clouqueur, and W. D. Grover. On the availability and capacity requirements of shared backup path-protected mesh networks. *SPIE Optical Networks Magazine*, 4(6): 29–44, 2003.

[97] J. Doucette and W. D. Grover. Comparison of mesh protection and restoration schemes and the dependency on graph connectivity. In *Proceedings 3rd International Workshop on the Design of Reliable Communication Networks (DRCN)*, Budapest, Hungary, October 2001.

[98] J. Doucette and W. D. Grover. Maintenance-immune optical mesh network design. In *Proceedings of the 18th Annual National Fiber Optic Engineers Conference (NFOEC)*, pages 2049–2061, Dallas, TX, September 2002.

[99] J. Doucette, W. D. Grover, and R. Martens. Modularity and economy-of-scale effects in the optimal design of mesh-restorable networks. In *Proc. of IEEE Canadian Conference on Electrical and Computer Engineering (CCECE)*, pages 226–231, Edmonton, AB, Canada, May 1999.

[100] R. Doverspike *et al.* Fast restoration in a mesh network of optical cross-connects. In *Proc. IEEE/OSA Optical Fiber Commun. Conf. (OFC)*, San Diego, CA, February 1999.

[101] R. Doverspike, S. Phillips, and J. Westbrook. Transport network architectures in an IP world. In *Proc. IEEE Infocom.*, pages 305–314, Tel Aviv, Israel, March 2000.

[102] R. Doverspike and J. Yates. Challenges for MPLS in optical network restoration. *IEEE Communications Mag.*, 39(2): 89–96, 2001.

[103] S. E. Dreyfus. An appraisal of some shortest-path algorithms. *Operations Research*, 17(3): 395–412, 1969.

[104] Z. Dziong, S. Kasera, and R. Nagarajan. Efficient capacity sharing in path restoration schemes for meshed optical networks. In *Proceedings of the 18th Annual National Fiber Optic Engineers Conference (NFOEC)*, Dallas, TX, September 2002.

[105] G. Ellinas. Fault restoration in optical networks: General methodology and implementation. PhD thesis, Columbia University, New York, NY, 1998.

[106] G. Ellinas *et al.* Restoration in layered architectures with a WDM mesh optical layer. *Annual Review of Communications for the International Engineering Consortium (IEC)*, vol.55, 2002.

[107] G. Ellinas *et al.* Routing and restoration architectures in mesh optical networks. *SPIE Optical Networks Magazine*, 4(1): 91–106, 2003.

[108] G. Ellinas *et al.* Transparent optical switches: Technology issues and challenges. *Annual Review of Communications for the International Engineering Consortium (IEC)*, vol.56, November 2003.

[109] G. Ellinas *et al.* Network control and management challenges in opaque networks utilizing transparent optical switches. *IEEE Communications Mag.*, 42(2): S16–S24, 2004.

[110] G. Ellinas, A. Hailemariam, and T. Stern. Optical switch failure protection in mesh networks with arbitrary non-planar architectures. In *Proc. IEEE Globecom.*, Rio de Janeiro, Brazil, December 1999.

[111] G. Ellinas, A. Hailemariam, and T. Stern. Protection of a priority connection from an optical switch failure in mesh networks with planar topologies. In *Proc. IEEE/OSA Optical Fiber Commun. Conf. (OFC)*, San Diego, CA, February 1999.

[112] G. Ellinas, S. Rong, A. Hailemariam, and T. Stern. Protection cycle covers in optical networks with arbitrary mesh topologies. In *Proc. IEEE/OSA Optical Fiber Commun. Conf. (OFC)*, Baltimore, MD, March 2000.

[113] G. Ellinas, T. Stern, and A. Hailemariam. Protection cycles in WDM mesh networks. *IEEE J. Select. Areas Commun.*, 18(10): 1924–1937, 2000.

[114] G. Ellinas and T. E. Stern. Automatic protection switching for link failures in optical networks with bi-directional links. In *Proc. IEEE Globecom.*, pages 152–156, London, UK, November 1996.

[115] A. F. Elrefaie. Self-healing WDM ring networks with an all-optical protection path. In *Proc. IEEE/OSA Optical Fiber Commun. Conf. (OFC)*, San Jose, CA, pages 255–256, February 1992.

[116] D. Eppstein. Finding the K shortest paths. *SIAM J. Comput.*, 28(2): 652–673, 1998.

[117] S. G. Finn, M. Medard, and R. A. Barry. A new algorithm for bi-directional link self-healing for arbitrary redundant networks. In *Proc. IEEE/OSA Optical Fiber Commun. Conf. (OFC)*, San Jose, CA, February 1998.

[118] S. G. Finn, M. M. Medard, and R. A. Barry. A novel approach to automatic protection switching using trees. In *Proc. IEEE Int. Conf. Commun. (ICC)*, pages 272–276, Montreal, Canada, June 1997.

[119] G.S. Fishman. *Monte Carlo: Concepts, Algorithms, and Applications*. Springer-Verlag, New York, NY, 1995.

[120] H. Fleischner. *Eulerian Graphs and Related Topics Part I: Vol. 1*, volume 45 of *Annals of Discrete Mathematics*. Elsevier Science Publishers B.V., Amsterdam, The Netherlands, 1990.

[121] J. E. Ford and J. Walker. Dynamic spectral power equalization using micro-mechanics. *IEEE Photonics Technology Letters*, 10(10): 1440–1442, 1998.

[122] L. R. Ford and D. R. Fulkerson. *Flows in Networks*. Princeton University Press, 1962.

[123] B. Forst and W. Grover. Testing the accuracy of analytical estimates of spare capacity in protected mesh networks. *J. Optical Networks*, 5(10): 715–738, 2006.

[124] S. French, J-F. Labourdette, K. Bala, and P. Miller-Pittman. Efficient network switching hierarchy. In *Proceedings of the 18th Annual National Fiber Optic Engineers Conference (NFOEC)*, Dallas, TX, September 2002.

[125] A. Fumagalli *et al.* Survivable networks based on optimal routing and WDM self-healing rings. In *Proc. IEEE Infocom.*, volume 2, New York, NY, March 1999.

[126] A. Fumagalli and L. Valacrenghi. IP restoration vs WDM protection: Is there an optimal choice? *IEEE Network Mag.*, 14(6): 34–41, 2000.

[127] L. M. Gardner, M. Heydari, J. Shah, I. H. Sudborough, I. G. Tollis, and C. Xia. Techniques for finding ring covers in survivable networks. In *Proc. IEEE Globecom.*, pages 1862–1866, San Francisco, CA, November 1994.

[128] M. R. Garey and D. S. Johnson. *Computers and Intractability: A Guide to NP-Completeness.* W.H. Freeman, New York, 1979.

[129] A. Gencata and B. Mukherjee. Virtual-topology adaptation for WDM mesh networks under dynamic traffic. In *Proc. IEEE Infocom.*, New York, NY, June 2002.

[130] K. C. Glossbrenner. Availability and reliability of switched services. *IEEE Communications Mag.*, 31(6): 28–33, 1993.

[131] L. Goddyn. *Cycles in Graphs*, chapter: A girth requirement for the cycle double cover conjecture. Elsevier Science Publishers BV, 1985.

[132] C. C. Gotlieb and D. G. Corneil. Algorithms for finding a fundamental set of cycles for an undirected linear graph. *Comm. ACM*, 10(12): 780–783, 1967.

[133] W. Grover. The protected working capacity envelope concept: An alternate paradigm for automated service provisioning. *IEEE Communications Mag.*, pages 62–69, January 2004.

[134] W. Grover and A. Kodian. Failure-independent path protection with *p*-cycles: Efficient, fast and simple protection for transparent optical networks. In *Proceedings of the 7th International Conference on Transparent Optical Networks*, pages 363–369, Barcelona, Spain, July 2005.

[135] W. D. Grover. The SelfHealing network: A fast distributed restoration technique for networks using digital crossconnect machines. In *Proc. IEEE Globecom.*, pages 1090–1095, Tokyo, Japan, November 1987.

[136] W. D. Grover. SelfHealing networks: A distributed algorithm for K-shortest link-disjoint paths in a multi-graph with applications in real time network restoration. PhD thesis, The University of Alberta, 1989.

[137] W. D. Grover. Case studies of survivable ring, mesh and mesh-arc hybrid networks. In *Proc. IEEE Globecom.*, pages 633–638, Orlando, FL, December 1992.

[138] W. D. Grover. *Telecommunications Network Management into the 21st Century*, chapter: Distributed restoration of the transport network. IEEE Press, New York, 1994.

[139] W. D. Grover. *Mesh-Based Survivable Networks: Options and Strategies for Optical, MPLS, SONET and ATM Networking*. Prentice-Hall, Englewood Cliffs, NJ, 2004.

[140] W. D. Grover, T. D. Bilodeau, and B. D. Venables. Near optimal spare capacity planning in a mesh restorable network. In *Proc. IEEE Globecom.*, pages 2007–2012, Phoenix, AZ, December 1991.

[141] W. D. Grover and M. Clouqueur. Span-restorable mesh networks with multiple quality of protection (QoP) service-classes. In *Proceedings of International Conference on Optical Communications and Networks (ICOCN)*, pages 321–323, Singapore, November 2002.

[142] W. D. Grover, B. D. Venables, J. H. Sandham, and A. F. Milne. Performance studies of a SelfHealing network protocol in Telecom Canada long haul networks. In *Proc. IEEE Globecom.*, pages 452–458, San Diego, CA, December 1990.

[143] W.D. Grover, M.Clouqueur, and T. Bach. Quantifying and managing the influence of maintenance actions on the survivability of mesh-restorable networks. In *Proceedings of the 17th Annual National Fiber Optic Engineers Conference (NFOEC)*, pages 1514–1525, Baltimore, MD, July 2001.

[144] W.D. Grover and D. Stamatelakis. Cycle-oriented distributed preconfiguration: Ring-like speed with mesh-like capacity for self-planning network restoration. In *Proc. IEEE Int. Conf. Commun. (ICC)*, pages 537–543, Atlanta, GA, June 1998.

[145] L. Guo *et al.* Dynamic shared-path protection based on SRLG constraints in WDM mesh networks. In *Proceedings of the IEEE International Conference on Communications, Circuits and Systems*, pages 643–646, Chengdu, China, June 2004.

[146] A. Hadjiantonis, A. Khalil, G. Ellinas, and M. Ali. A novel restoration scheme for next generation WDM-based IP backbone networks. In *Proc. LEOS Annual Meeting*, Tucson, Arizona, October 2003.

[147] A. Hailemariam. Island-based restoration in mesh optical networks. PhD thesis, Columbia University, New York, NY, 2003.

[148] A. Hailemariam, G. Ellinas, and T. Stern. Localized restoration in mesh optical networks. In *Proc. IEEE/OSA Optical Fiber Commun. Conf. (OFC)*, Los Angeles, CA, February 2004.

[149] R. D. Hall and S. Whitt. Protection of SONET based networks. In *Proc. IEEE Globecom.*, pages 821–825, Dallas, TX, November 1989.

[150] G. Held. On the road to OC-768. *IT Professional*, 3(2): 46–48, March–April 2001.

[151] A. Herschtal and M. Herzberg. Dynamic capacity allocation and optimal rearrangement for reconfigurable networks. In *Proc. IEEE Globecom.*, pages 946–951, Singapore, 1990.

[152] M. Herzberg. A decomposition approach to assign spare channels in self-healing networks. In *Proc. IEEE Globecom.*, Houston, TX, December 1993.

[153] M. Herzberg and S. Bye. An optimal spare-capacity assignment model for survivable networks with hop limits. In *Proc. IEEE Globecom.*, pages 1601–1607, San Francisco, CA, 1994.

[154] H. S. Hinton. *An Introduction to Photonic Switching Fabrics*. Plenum Press, New York, 1993.

[155] G. Hjalmtysson, J. Yates, S. Chaudhuri, and A. Greenberg. Smart routers simple optics: An architecture for the optical internet. *IEEE/OSA J. Lightwave Technology*, 18(12): 161–167, 2000.

[156] P. Ho and H. T. Mouftah. A framework for service-guaranteed shared protection in WDM mesh networks. *IEEE Communications Mag.*, pages 97–103, February 2002.

[157] P.-H. Ho and H. T. Mouftah. A framework of a survivable optical internet using short leap shared protection (SLSP). In *Proc. IEEE Workshop on High Performance Switching and Routing (HPSR)*, Dallas, TX, May 2001.

[158] P. H. Ho and H. T. Mouftah. Allocation of protection domains in dynamic WDM mesh networks. In *Proc. IEEE Int. Conf. Commun. (ICC)*, Seattle, WA, May 2003.

[159] P. H. Ho and H. T. Mouftah. Spare capacity allocation for WDM mesh networks with partial wavelength conversion capacity. In *Proc. IEEE High Performance Switching and Routing*, Turin, Italy, June 2003.

[160] P. Humblet. The direction of optical technology in the metro area. In *Proc. IEEE/OSA Optical Fiber Commun. Conf. (OFC)*, Anaheim, CA, March 2001.

[161] Ilog Inc. CPLEX. http://www.ilog.com/products/cplex/.

[162] R. R. Iraschko, W. D. Grover, and M. H. MacGregor. A distributed real time path restoration protocol with performance close to centralized multi-commodity max flow. In *Proceedings of the First International Workshop on Design of Reliable Communication Networks (DRCN)*, Bruges, Belgium, May 1998.

[163] R. R. Iraschko, M. H. MacGregor, and W. D. Grover. Optimal capacity placement for path restoration in mesh survivable networks. In *Proc. IEEE Int. Conf. Commun. (ICC)*, pages 1568–1574, Dallas, TX, June 1996.

[164] R. R. Iraschko, M. H. MacGregor, and W. D. Grover. Optimal capacity placement for path restoration in STM or ATM mesh-survivable networks. *IEEE/ACM Trans. Networking*, 6(3): 325–336, 1998.

[165] F. Jaeger. Flows and generalized coloring theorems in graphs. *J. Combinatorial Theory*, 26(2): 205–216, 1979.

[166] F. Jaeger. *Cycles in Graphs*, chapter: A survey of the cycle double cover conjecture. Elsevier Science Publishers BV, 1985.

[167] F. Jaeger and T. Swart. *Combinatorics 79*, chapter: Conjecture 1. Elsevier Science Publishers BV, 1980.

[168] J. Jaffe. Algorithm for finding path with multiple constraints. *Networks*, 14: 95–116, 1984.

[169] C. Janczewski, D. Jeske, K. Mrti, R. Nagarajan, and A. Sampath. Cost, restoration, performance and reliability of optical networks with shared and dedicated protection schemes. In *Proceedings of the 18th Annual National Fiber Optic Engineers Conference (NFOEC)*, Dallas, TX, September 2002.

[170] S. Johansson. A transport network involving a reconfigurable WDM network layer – A European demonstration. *IEEE/OSA J. Lightwave Technology*, 14(6): 1341–1348, 1996.

[171] D. Johnson, G. N. Brown, S. L. Beggs, C. P. Botham, I. Hawker, R. S. K. Chng, M. C. Sinclair, and M. J. O'Mahony. Distributed restoration strategies in telecommunications networks. In *Proc. IEEE Int. Conf. Commun. (ICC)*, New Orleans, LA, May 1994.

[172] A. Jukan. *QoS-based Wavelength Routing in Multi-Service WDM Networks*. Springer-Verlag, New York, 2001.

[173] A. Jukan and G. Franzl. Path selection methods with multiple constraints in service-guaranteed WDM networks. *IEEE/ACM Trans. Networking*, 12(1): 59–72, 2004.

[174] N. Katoh, T. Ibaraki, and H. Mine. An efficient algorithm for k shortest simple paths. *Networks*, 12(4): 411–427, 1982.

[175] D. Katz *et al. Traffic Engineering Extensions to OSPF Version 2*. Technical Report RFC 3630, Internet Engineering Task Force (IETF), September 2003.

[176] J. Kirkpatrick. Private communication, December 2006.

[177] H. Kobrinski and M. Azuma. Distributed control algorithms for dynamic restoration in DCS mesh networks: Performance evaluation. In *Proc. IEEE Globecom.*, pages 1584–1588, Houston, TX, November 1993.

[178] M. Kodialam and T.V. Lakshman. Integrated dynamic IP and wavelength routing in IP over WDM networks. In *Proc. IEEE Infocom.*, pages 358–366, Anchorage, Alaska, 2001.

[179] M. Kodialan and T.V. Lakshman. Dynamic routing of bandwidth guaranteed tunnels with restoration. In *Proc. IEEE Infocom.*, Tel Aviv, Israel, March 2000.

[180] H. Komine, T. Chujo, T. Ogura, K. Miyazaki, and T. Soejima. A distributed restoration algorithm for multiple-link and node failures of transport networks. In *Proc. IEEE Globecom.*, pages 459–463, San Diego, December 1990.

[181] K. Kompella *et al.* *OSPF Extensions in Support of Generalized MPLS.* Technical Report RFC 4203, Internet Engineering Task Force (IETF), October 2005.

[182] K. Kompella *et al.* *Routing Extensions in Support of Generalized MPLS.* Technical Report RFC 4202, Internet Engineering Task Force (IETF), October 2005.

[183] S. K. Korotky. Network global expectation model: A statistical formalism for quickly quantifying network needs and costs. *IEEE/OSA J. Lightwave Technology*, 22(3): 703–722, 2004.

[184] S. K. Korotky. An overview of the network global expectation model. In *Proc. IEEE/OSA Optical Fiber Commun. Conf. (OFC)*, Los Angeles, CA, March 2004.

[185] S. K. Korotky. Analysis and optimization of two-tier networks with application to optical transport backbones. *J. Optical Network*, 4(1): 45–63, 2005.

[186] J. Labourdette, E. Bouillet, and S. Chaudhuri. Role of optical network and spare router design in resilient IP backbone architectures. In *Proc. Design of Reliable Communication Networks (DRCN)*, Banff, Canada, October 2003.

[187] J. Labourdette, E. Bouillet, R. Ramamurthy, G. Ellinas, S. Chaudhuri, and K. Bala. Routing strategies for capacity-efficient and fast-restorable mesh optical networks. *Photonic Network Communications*, 4(3/4): 219–235, 2002.

[188] J. Labourdette *et al.* Fast approximate dimensioning and performance analysis of mesh optical networks. In *Proc. Design of Reliable Communication Networks (DRCN)*, Banff, Canada, October 2003.

[189] J. Labourdette *et al.* Layered architecture for scalability in core mesh optical networks. In *Proceedings of the 19th Annual National Fiber Optic Engineers Conference (NFOEC)*, Orlando, FL, September 2003.

[190] J. Labourdette *et al.* Fast approximate dimensioning and performance analysis of mesh optical networks. *IEEE/ACM Trans. Networking*, 13(4): 906–917, 2005.

[191] J.-F. P. Labourdette, G. W. Hart, and A. S. Acampora. Branch-exchange sequences for reconfiguration of lightwave networks. *IEEE Trans. Commun.*, 42(10): 2822–2832, 1994.

[192] W. Lai *et al.* *Network Hierarchy and Multilayer Survivability*, Internet Engineering Task Force (IETF) RFC 3386, November 2002.

[193] J. Lang *et al.* *Link Management Protocol (LMP).* Technical Report RFC 4204, Internet Engineering Task Force (IETF), October 2005.

[194] A. Lardies, R. Gupta, and R. Patterson. Traffic grooming in a multi-layer network. *SPIE Optical Networks Magazine*, 2(3): 91–99, 2001.

[195] E. L. Lawler. A procedure for computing the K best solutions to discrete optimization problems and its application to the shortest path problem. *Management Science*, 18(7): 401–405, 1972.

[196] D. Leung, S. Arakawa, M. Murata, and W. D. Grover. Reoptimization strategies to maximize traffic-carrying readiness in WDM survivable mesh networks. In *Proc. IEEE/OSA Optical Fiber Commun. Conf. (OFC)*, Anaheim, CA, March 2005.

[197] C. Li *et al*. Gain equalization in metropolitan and wide area optical networks using optical amplifiers. In *Proc. IEEE Infocom.*, pages 130–137, Toronto, Canada, July 1994.

[198] G. Li *et al*. Efficient distributed path selection for shared restoration connections. In *Proc. IEEE Infocom.*, New York, NY, July 2002.

[199] H. Liu. *OSPF-TE Extensions in Support of Shared Mesh Restoration*. Technical Report draft-liu-gmpls-ospf-restoration-00.txt, Internet Engineering Task Force (IETF), October 2002.

[200] H. Liu *et al*. Distributed route computation algorithms and dynamic provisioning in intelligent mesh optical networks. In *Proc. IEEE Int. Conf. Commun. (ICC)*, Anchorage, Alaska, May 2003.

[201] Y. Liu, D. Tipper, and P. Siripongwutikorn. Approximating optimal spare capacity allocation by successive survivable routing. In *Proc. IEEE Infocom.*, Anchorage, AL, April 2001.

[202] Y. Liu, D. Tipper, and P. Siripongwutikorn. Approximating optimal spare capacity allocation by successive survivable routing. *IEEE/ACM Trans. Networking*, 13(1): 198–211, 2005.

[203] J. Livas. Optical transmission evolution: From digital to analog to ? Network tradeoffs between optical transparency and reduced regeneration cost. *IEEE/OSA J. Lightwave Technology*, 23(1): 219–224, 2005.

[204] S. Lumetta and M. Medard. Towards a deeper understanding of link restoration algorithms for mesh networks. In *Proc. IEEE Infocom.*, Anchorage, Alaska, April 2001.

[205] X. Luo and B. Wang. Diverse routing in WDM optical networks with shared risk link group (SRLG) failures. In *Proceedings 5th International Workshop on Design of Reliable Communication Networks (DRCN)*, Ischia, Italy, October 2005.

[206] M. H. Macgregor and W. D. Grover. Optimized k-shortest-paths algorithm for facility restoration. *Software–Practice and Experience*, 24(9): 823–834, 1994.

[207] N. Madamopoulos *et al*. Design, transport performance study and engineering of a 11 Tb/s US metro network. In *Proc. IEEE/OSA Optical Fiber Commun. Conf. (OFC)*, pages 444–445, Anaheim, CA, March 2002.

[208] S. Makam *et al*. *Framework for MPLS Based Recovery*. Technical Report RFC 3469, Internet Engineering Task Force (IETF), February 2003.

[209] J. Manchester, P. Bonenfant, and C. Newton. The evolution of transport network survivability. *IEEE Communications Mag.*, pages 44–51, August 1999.

[210] E. Mannie *et al*. *Generalized Multi-Protocol Label Switching (GMPLS) Architecture*. Technical Report RFC 3945, Internet Engineering Task Force (IETF), October 2004.

[211] P. Mateti and N. Deo. On algorithms for enumerating all circuits of a graph. *SIAM J. Comput.*, 5(1): 90–99, 1976.

[212] A. McGuire, S. Mirza, and D. Freeland. Application of control plane technology to dynamic configuration management. *IEEE Communications Mag.*, pages 94–99, September 2001.

[213] M. Medard, R. A. Barry, S. G. Finn, W. He, and S. Lumetta. Generalized loopback recovery in optical mesh networks. *IEEE/ACM Trans. Networking*, 10(1): 153–164, 2002.

[214] M. Medard, S. G. Finn, R. A. Barry, and R. G. Gallager. Redundant trees for pre-planned recovery in arbitrary vertex-redundant or edge-redundant graphs. *IEEE/ACM Trans. Networking*, 7(5): 641–652, 1999.

[215] D. Medhi. Multi-hour, multi-traffic class network design for virtual path-based dynamically reconfigurable wide-area ATM networks. *IEEE/ACM Trans. Networking*, 3(6): 809–818, 1995.

[216] E. Modiano and P. Lin. Traffic grooming in WDM networks. *IEEE Communications Mag.*, 39(7): 124–129, 2001.

[217] A. Morea *et al*. Impact of the reach of WDM systems and traffic volume on the network resources and cost of translucent optical transport networks. In *Proceedings of the International Conference on Transparent Optical Networks (ICTON)*, pages 65–68, Wroclaw, Poland, July 2004.

[218] A. Morea and J. Poirrier. A critical analysis of the possible cost savings of translucent networks. In *Proceedings of the 5th International Workshop on Design of Reliable Communication Networks (DRCN)*, Ischia, Italy, October 2005.

[219] J. Moy *et al*. OSPF Version 2. Technical Report RFC 2328, Internet Engineering Task Force (IETF), April 1998.

[220] B. Mukherjee. WDM optical communication networks: Progress and challenges. *IEEE J. Select. Areas Commun.*, 18(10): 1810–1824, 2000.

[221] B. Mukherjee. *Optical WDM Networks*. Springer-Verlag, New York, 2006.

[222] R. Nagarajan *et al*. Large scale photonic integrated circuits. *IEEE Journal of Sel. Top. In Quant. Elec.*, 11(1): 50–65, 2005.

[223] T. Naito *et al*. 1 Terabits/s WDM transmission over 10,000 Km. In *Proceedings European Conference on Optical Communications (ECOC)*, Nice, France, September 1999.

[224] G. L. Nemhauser and L. A. Wolsey. *Integer and Combinatorial Optimization*. John Wiley & Sons, New York, 1988.

[225] T. Nishizeki and N. Chiba. *Planar Graphs: Theory and Algorithms*, volume 32 of *Annals of Discrete Mathematics*. Elsevier Science Publishers BV, Amsterdam, The Netherlands, 1988.

[226] K. Noguchi *et al*. The first field trial of a wavelength routing WDM full-mesh network system (AWG-STAR) in a metropolitan/local area. In *Proc. IEEE/OSA Optical Fiber Commun. Conf. (OFC)*, pages 611–613, Atlanta, GA, February 2003.

[227] T. H. Noh, D. R. Vaman, and X. Gu. Reconfiguration for service and self-healing in ATM networks based on virtual paths. *Computer Networks*, 29(16): 1857–1867, 1997.

[228] E. Oki *et al*. A disjoint path selection scheme with shared risk link groups in GMPLS networks. *IEEE Communications Letters*, 6(9): 406–408, 2002.

[229] C. Olzsewski *et al*. Two-tier network economics. In *Proceedings of the 18th Annual National Fiber Optic Engineers Conference (NFOEC)*, Dallas, TX, September 2002.

[230] M. J. O'Mahony *et al*. The design of a European optical network. *IEEE/OSA J. Lightwave Technology*, 13(5): 817–828, 1995.

[231] *Optical Network Magazine*. Special Issue on Telecommunications Grooming. 2(3), May/June 2001.

[232] C. Ou, H. Zang, and B. Mukherjee. Sub-path protection for scalability and fast recovery in WDM mesh networks. *Proc. IEEE/OSA Optical Fiber Commun. Conf. (OFC)*, Anaheim, CA, March 19–22, 2001.

[233] J. C. Palais. *Fiber-Optic Communications*. Prentice-Hall, Englewood Cliffs, NJ, 2004.

[234] X. Pan and G. Xiao. Algorithms for the diverse routing problem in WDM networks with shared risk link groups. In *Proceedings of the 9th IEEE International Conference on Communications Systems*, pages 381–385, Singapore, September 2004.

[235] D. Papadimitriou. *Analysis of Generalized MPLS-based Recovery Mechanisms (including Protection and Restoration)*. Technical Report RFC 4428, Internet Engineering Task Force (IETF), March 2006.

[236] K. Paton. An algorithm for finding a fundamental set of cycles for a graph. *Comm. ACM*, 12(9): 514–518, 1969.

[237] P. Peloso, D. Penninckx, M. Prunaire, and L. Noirie. Optical transparency of a heterogeneous pan-European network. *IEEE/OSA J. Lightwave Technology*, 22(11): 242–248, 2004.

[238] M. Pioro and D. Medhi. *Routing, Flow, and Capacity Design in Communication and Computer Networks*. Morgan Kaufmann Publishers, San Mateo, California, 2005.

[239] A. Pratt *et al.* 40x10.7 Gb/s DWDM transmission over a meshed ULH network with dynamically reconfigurable optical crossconnects. In *Proc. IEEE/OSA Optical Fiber Commun. Conf. (OFC)*, Atlanta, GA, March 2003.

[240] C. Qiao *et al.* Distributed partial information management (DPIM) schemes for survivable networks – Part I. In *Proc. IEEE Infocom.*, New York, NY, July 2002.

[241] C. Qiao *et al.* Novel models for efficient shared-path protection. In *Proc. IEEE/OSA Optical Fiber Commun. Conf. (OFC)*, Anaheim, CA, March 2002.

[242] Y. Qin, L. Mason, and K. Jia. Study on a joint multiple layer restoration scheme for IP over WDM networks. *IEEE Network Mag.*, pages 43–48, March/April 2003.

[243] B. Rajagopalan *et al. IP over Optical Networks: A Framework*. Technical Report RFC 3717, Internet Engineering Task Force (IETF), March 2004.

[244] B. Rajagopalan, D. Pendarakis, D. Saha, S. Ramamurthy, and K. Bala. IP over optical networks: Architectural aspects. *IEEE Communications Mag.*, pages 94–102, September 2000.

[245] B. Ramamurthy *et al.* Optimizing amplifier placements in a multi-wavelength optical LAN/MAN: The equally-powered-wavelengths case. *IEEE/OSA J. Lightwave Technology*, 16(9): 1560–1569, 1998.

[246] B. Ramamurthy *et al.* Impact of transmission impairments on the teletraffic performance of wavelength-routed optical networks. *IEEE/OSA J. Lightwave Technology*, 17(10): 1713–1723, 1999.

[247] B. Ramamurthy *et al.* Transparent vs. opaque vs. translucent wavelength-routed optical networks. In *Proc. IEEE/OSA Optical Fiber Commun. Conf. (OFC)*, pages 59–61, San Diego, CA, February 1999.

[248] B. Ramamurthy *et al*. Translucent optical WDM networks for the next-generation backbone networks. In *Proc. IEEE Globecom.*, San Antonio, TX, November 2001.

[249] R. Ramamurthy, A. Akyamaç, J. Labourdette, and S. Chaudhuri. Pre-emptive reprovisioning in mesh optical networks. In *Proc. IEEE/OSA Optical Fiber Commun. Conf. (OFC)*, Atlanta, GA, March 2003.

[250] R. Ramamurthy, Z. Bogdanowicz, S. Samieian, D. Saha, B. Rajagopalan, S. Sengupta, S. Chaudhuri, and K. Bala. Capacity performance of dynamic provisioning in optical networks. *IEEE/OSA J. Lightwave Technology*, 19(1): 40–48, 2001.

[251] R. Ramamurthy *et al*. Comparison of centralized and distributed provisioning of lightpaths in optical networks. In *Proc. IEEE/OSA Optical Fiber Commun. Conf. (OFC)*, Anaheim, CA, March 2001.

[252] R. Ramamurthy *et al*. Routing lightpaths in optical mesh networks with express links. In *Proc. IEEE/OSA Optical Fiber Commun. Conf. (OFC)*, Anaheim, CA, March 2002.

[253] R. Ramamurthy *et al*. Limiting sharing on protection channels in mesh optical networks. In *Proc. IEEE/OSA Optical Fiber Commun. Conf. (OFC)*, Atlanta, GA, March 2003.

[254] R. Ramamurthy, J. Labourdette, and E. Bouillet. Limitations of scaling switching capacity by interconnecting multiple switches. In *Proceedings of the Annual National Fiber Optic Engineers Conference (NFOEC)*, Anaheim, CA, March 2006.

[255] S. Ramamurthy and B. Mukherjee. Survivable WDM mesh networks. Part I Protection. In *Proc. IEEE Infocom.*, New York, NY, March 1999.

[256] S. Ramamurthy and B. Mukherjee. Survivable WDM mesh networks. Part II Restoration. In *Proc. IEEE Int. Conf. Commun. (ICC)*, Vancouver, BC, Canada, June 1999.

[257] R. Ramaswami and K. Sivarajan. *Optical Networks: A Practical Perspective*, 2nd Edition. Morgan Kaufmann Publishers, San Mateo, California, 2002.

[258] R. Ramaswami and K. N. Sivarajan. Routing and wavelength assignment in all-optical networks. *IEEE/ACM Trans. Networking*, 3(5): 489–500, 1995.

[259] R. Ranganathan, L. Blair, and J. Berthold. Architectural implications of core grooming in a 46-node USA optical network. In *Proc. IEEE/OSA Optical Fiber Commun. Conf. (OFC)*, pages 498–499, Anaheim, CA, March 2002.

[260] R. C. Read. *Theory of Graphs*, chapter: Graph theory algorithms. Academic Press, 1969.

[261] G. N. Rouskas and M. Ammar. Dynamic reconfiguration in multihop WDM networks. *Journal of High Speed Networks*, 4(3): 221–238, 1995.

[262] L. Sahasrabuddh, S. Ramamurthy, and B. Mukherjee. Fault management in IP-over-WDM networks: WDM protection versus IP restoration. *IEEE J. Select. Areas Commun.*, 20(1): 21–33, 2002.

[263] G. Sahin and M. Azizoglu. Optical layer survivability: Single service-class case. *Proceedings of SPIE*, 4233: 267–278, 2000.

[264] G. Sahin and M. Azizoglu. Optical layer survivability for single and multiple service classes. *Journal of High Speed Networks*, 10(2): 91–108, 2001.

[265] H. Sakauchi, Y. Nishimura, and S. Hasegawa. A Self-Healing network with an economical spare-channel assignment. In *Proc. IEEE Globecom.*, pages 438–443, San Diego, CA, December 1990.

[266] A. A. M. Saleh. Transparent optical networking in backbone networks. In *Proc. IEEE/OSA Optical Fiber Commun. Conf. (OFC)*, pages 62–64, Baltimore, MD, February 2000.

[267] A. A. M. Saleh. All-optical networking in metro, regional, and backbone networks. In *Proc. IEEE/LEOS Summer Topical Meeting*, Quebec, Canada, July 2002.

[268] A. A. M. Saleh. Defining all-optical networking and assessing its benefits in metro, regional and backbone networks. In *Proc. IEEE/OSA Optical Fiber Commun. Conf. (OFC)*, pages 410–411, Atlanta, GA, February 2003.

[269] A.A.M. Saleh and J. M. Simmons. Architectural principles of optical regional and metropolitan access networks. *IEEE/OSA J. Lightwave Technology*, 17(12): 2431–2448, 1999.

[270] K. Sato and S. Okamoto. Photonic transport technologies to create robust backbone networks. *IEEE Communications Mag.*, pages 78–87, August 1999.

[271] D. A. Schupke, W. D. Grover, and M. Clouqueur. Strategies for enhanced dual failure restorability with static or reconfigurable *p*-cycle networks. In *Proc. IEEE Int. Conf. Commun. (ICC)*, Paris, France, June 2004.

[272] D. A. Schupke, C. Gruber, and A. Autenrieth. Optimal configuration of *p*-cycles in WDM networks. In *Proc. IEEE Int. Conf. Commun. (ICC)*, pages 2761–2765, New York, NY, June 2002.

[273] P. Sebos, J. Yates, G. Hjalmtysson, and A. Greenberg. Auto-discovery of shared risk link groups. In *Proc. IEEE/OSA Optical Fiber Commun. Conf. (OFC)*, pages WDD3/1–WDD3/3, Anaheim, CA, March 2001.

[274] P. Sebos, J. Yates, D. Rubenstein, and A. Greenberg. Effectiveness of shared risk link group auto-discovery in optical networks. In *Proc. IEEE/OSA Optical Fiber Commun. Conf. (OFC)*, Anaheim, CA, March 2002.

[275] S. Sengupta, S. Chaudhuri, and D. Saha. Reliability of optical mesh and ring networks. In *OptoElectronics and Communications Conference (OECC)*, Sydney, Australia, July 2001.

[276] S. Sengupta, V. Kumar, and D. Saha. Switched optical backbone for cost-effective scalable IP networks. *IEEE Communications Mag.*, pages 60–70, June 2003.

[277] G. Shen *et al.* Sparse placement of electronic switching nodes for low blocking in translucent optical networks. *OSA J. Optical Network*, 1(12): 424–441, 2002.

[278] G. Shen and W. D. Grover. Extending the *p*-cycle concept to path segment protection for span and node failure recovery. *IEEE J. Select. Areas Commun.*, 21(8): 1306–1319, 2003.

[279] L. Shen, X. Yang, and B. Ramamurthy. A load-balancing shared-protection-path reconfiguration approach in WDM wavelength-routed networks. In *Proc. IEEE/OSA Optical Fiber Commun. Conf. (OFC)*, Anaheim, CA, February 2004.

[280] D. Shier. Computational experience with an algorithm for finding the k shortest paths in a network. *Journal of Research of the NBS*, 78: 139–164, 1974.

[281] D. Shier. Interactive methods for determining the k shortest paths in a network. *Networks*, 6: 151–159, 1976.

[282] D. Shier. On algorithms for finding the k shortest paths in a network. *Networks*, 9: 195–214, 1979.

[283] J. M. Simmons. On determining the optimal optical reach for a long-haul network. *IEEE/OSA J. Lightwave Technology*, 23(3): 1039–1048, 2005.

[284] J. B. Slevinsky, W. D. Grover, and M. H. MacGregor. An algorithm for survivable network design employing multiple self-healing rings. In *Proc. IEEE Globecom.*, pages 1568–1573, Houston, TX, November 1993.

[285] M. Sridharan, A. K. Somani, and M. V. Salapaka. Approaches for capacity and revenue optimization in survivable WDM networks. *Journal of High Speed Networks*, 10(2): 109–125, 2001.

[286] D. Stamatelakis and W. D. Grover. IP layer restoration and network planning based on virtual protection cycles. *IEEE J. Select. Areas Commun.*, 18(10): 1938–1949, 2000.

[287] D. Stamatelakis and W. D. Grover. Theoretical underpinning for the efficiency of restorable networks using preconfigured cycles (*p*-cycles). *IEEE Trans. Commun.*, 48(8):1262–1265, 2000.

[288] American National Standards. ANSI T1.105 1988.

[289] T. E. Stern and K. Bala. *Multiwavelength Optical Networks: A Layered Approach*. Addison-Wesley, Reading, MA, 1999.

[290] J. Strand, A. Chiu, and R. Tkach. Issues for routing in the optical layer. *IEEE Communications Mag.*, pages 81–87, February 2001.

[291] K. Struyve and P. Demeester. Design of distributed restoration algorithms for ATM meshed networks. In *Proc. IEEE Third Symposium on Communications and Vehicular Technology in the Benelux*, Eindhoven, The Netherlands, October 1995.

[292] J. W. Suurballe and R. E. Tarjan. A quick method for finding shortest pairs of disjoint paths. *Networks*, 14: 325–336, 1984.

[293] J.W. Suurballe. Disjoint paths in a network. *Networks*, 14: 125–145, 1974.

[294] J. Tapolcai and P.-H. Ho. Linear formulation for segment shared protection. In *Proc. OptiComm*, Dallas, TX, October 2003.

[295] R. Tarjan. Depth-first search and linear graph algorithms. *SIAM J. Comput.*, 1(2): 146–160, 1972.

[296] R. E. Tarjan. *Data Structures and Network Algorithms*. Society for Industrial and Applied Mathematics, Philadelphia, PA, 1983.

[297] M. Tarsi. Semi-duality and the cycle double cover conjecture. *J. Combinatorial Theory*, 41(3): 332–340, 1986.

[298] S. Thiagarajan and A. Somani. Capacity fairness of WDM networks with grooming capabilities. *SPIE Optical Networks Magazine*, 2(3): 24–31, 2001.

[299] R. W. Tkach, E. L. Goldstein, J. A. Nagel, and J. L. Strand. Fundamental limits of optical transparency. In *Proc. IEEE/OSA Optical Fiber Commun. Conf. (OFC)*, San Jose, CA, February 1998.

[300] M. To and P. Neusy. Unavailability analysis of long-haul networks. *IEEE J. Select. Areas Commun.*, 12(1): 100–109, 1994.

[301] P. Toliver, R. Runser, J. Young, and J. Jackel. Experimental field trial of waveband switching and transmission in a transparent reconfigurable optical network. In *Proc. IEEE/OSA Optical Fiber Commun. Conf. (OFC)*, pages 783–784, Atlanta, GA, February 2003.

[302] I. Tomkos *et al.* 80x10.7Gb/s ultra long-haul (+4200 Km) DWDM network with reconfigurable broadcast and select OADMs. In *Proc. IEEE/OSA Optical Fiber Commun. Conf. (OFC)*, Anaheim, CA, March 2002.

[303] M. Tsurusawa, J. Matsuda, T. Otani, M. Daikoku, S. Aruga, and H. Tanaka. Field trial of 10-Gb/s resilient packet ring for a next-generation multi-service metropolitan area network. In *Proc. IEEE/OSA Optical Fiber Commun. Conf. (OFC)*, Anaheim, CA, March 2005.

[304] E. Varma, S. Sankaranarayanan, G. Newsome, Z. Lin, and H. Epstein. Architecting the services optical network. *IEEE Communications Mag.*, pages 80–87, September 2001.

[305] J-P. Vassuer, P. Demeester, and M. Pickavet. *Network Recovery: Protection and Restoration of Optical, SONET SDH, IP and MPLS*. Morgan Kaufmann Publishers, San Mateo, California, 2004.

[306] B. O. Venables, W. D. Grover, and M. H. MacGregor. Two strategies for spare capacity placement in mesh restorable networks. In *Proc. IEEE Int. Conf. Commun. (ICC)*, pages 267–271, Helsinki, Finland, 2001.

[307] R. Vodhanel *et al.* National scale WDM networking demonstration by the MONET consortium. In *Proc. IEEE/OSA Optical Fiber Commun. Conf. (OFC)*, Dallas, TX, February 1997.

[308] R. E. Wagner, R. C. Alferness, A. A. M. Saleh, and M. S. Goodman. MONET: Multiwavelength Optical Networking. *IEEE/OSA J. Lightwave Technology*, 14(6): 1349–1355, 1996.

[309] R. E. Wagner *et al.* The potential of optical layer networks. In *Proc. IEEE/OSA Optical Fiber Commun. Conf. (OFC)*, Anaheim, CA, March 2001.

[310] J. Walker. The future of MEMS in telecommunication networks. *J. Micromech. Microeng.*, 10(3): R1–R7, 2000.

[311] A. F. Wallace. Ultra long-haul DWDM: Network economics. In *Proc. IEEE/OSA Optical Fiber Commun. Conf. (OFC)*, Anaheim, CA, March 2001.

[312] Z. Wang and J. Crowcroft. Quality of service routing for supporting multimedia applications. *IEEE J. Select. Areas Commun.*, 14(7): 1228–1234, 1996.

[313] O. J. Wasem. An algorithm for designing rings for survivable fiber networks. *IEEE Trans. Reliability*, 40(4): 428–432, 1991.

[314] O. J. Wasem. Optimal topologies for survivable fiber optic networks using SONET self-healing rings. In *Proc. IEEE Globecom.*, pages 2032–2038, Phoenix, AZ, December 1991.

[315] O. J. Wasem, R. H. Cardwell, and T-H. Wu. Software for designing survivable SONET networks using self-healing rings. In *Proc. IEEE Int. Conf. Commun. (ICC)*, pages 425–431, Chicago, IL, June 1992.

[316] H. Wen *et al.* Dynamic RWA algorithms under shared-risk-link-group constraints. In *Proc. of the IEEE International Conference on Communications, Circuits and Systems and West Sino Expositions*, pages 871–875, Chengdu, China, June 2002.

[317] P. Wong. Network architectures in the metro DWDM environment. In *Proc. National Fiber Optics Engineers Conference (NFOEC)*, pages 588–596, Denver, CO, August 2000.

[318] L. A. Wrobel. *Disaster Recovery Planning for Telecommunications*. Artech House, Norwood, MA, 1990.

[319] T-H. Wu. *Fiber Network Service Survivability*. Artech House, Norwood, MA, 1992.

[320] T-H. Wu. Roles of optical components in survivable fiber networks. In *Proc. IEEE/OSA Optical Fiber Commun. Conf. (OFC)*, pages 252–253, San Jose, CA, February 1992.

[321] T-H. Wu and R. C. Lau. A class of self-healing ring architectures for SONET network applications. In *Proc. IEEE Globecom.*, pages 444–451, San Diego, CA, December 1990.

[322] Y. Xiong and L. G. Mason. Restoration strategies and spare capacity requirements in self-healing ATM networks. *IEEE/ACM Trans. Networking*, 7(1): 98–110, 1999.

[323] D. Xu, Y. Chen, Y. Xiong, C. Qiao, and X. He. On finding disjoint paths in single and dual link cost networks. In *Proc. IEEE Infocom.*, Hong Kong, March 2004.

[324] D. Xu, Y. Xiong, and C. Qiao. Protection with multi-segments (PROMISE) in networks with shared risk link groups (SRLG). In *40th Annual Allerton Conference on Communication, Control, and Computing*, Allerton, IL, October 2002.

[325] D. Xu, Y. Xiong, and C. Qiao. Novel algorithms for shared segment protection. *IEEE J. Select. Areas Commun.*, 21(8): 1320–1331, 2003.

[326] C. H. Yang and S. Hasegawa. FITNESS: Failure immunization technology for network service survivability. In *Proc. IEEE Globecom.*, pages 1549–1554, Hollywood, FL, November 1988.

[327] X. Yang and B. Ramamurthy. Dynamic routing in translucent WDM optical networks. In *Proc. IEEE Int. Conf. Commun. (ICC)*, pages 2796–2802, New York City, NY, April 2002.

[328] Y. Ye, C. Assi, S. Dixit, and M. Ali. A simple dynamic integrated provisioning/protection scheme in IP over WDM networks. *IEEE Communications Mag.*, pages 174–182, November 2001.

[329] Y. Ye, S. Dixit, and M. Ali. On joint protection/restoration in IP-centric DWDM-based optical transport networks. *IEEE Communications Mag.*, pages 174–183, June 2000.

[330] J. Y. Yen. Finding the K shortest loopless paths in a network. *Management Science*, 17(11): 712–716, 1971.

[331] C. Yihong *et al.* Metro network with reconfigurable optical add-drop module. In *IEEE/LEOS Summer Topical Meeting*, pages TuF3–25–TuF3–26, Quebec, Canada, July 2002.

[332] H. Zang *et al.* Path-protection routing and wavelength assignment (RWA) in WDM mesh networks under duct-layer constraints. *IEEE/ACM Trans. Networking*, 11(2): 248–258, 2003.

[333] H. Zeng, C. Huang, A. Vukovic, and J. M. Savoie. Transport performance of an all-optical metro WDM network based on dynamic all-optical switches. In *Proc. IEEE/OSA Optical Fiber Commun. Conf. (OFC)*, Anaheim, CA, March 2005.

[334] H. Zhu, H. Zang, K. Zhu, and B. Mukherjee. A novel, generic graph model for traffic grooming in heterogeneous WDM mesh networks. *IEEE/ACM Trans. Networking*, 11(2): 285–299, 2003.

[335] K. Zhu and B. Mukherjee. Traffic grooming in an optical WDM mesh network. *IEEE J. Select. Areas Commun.*, 20(1): 122–133, 2002.

[336] K. Zhu and B. Mukherjee. A review of traffic grooming in WDM optical networks: Architectures and challenges. *SPIE Optical Networks Magazine*, 4(2): 55–64, 2003.

Index

Path Routing in Mesh Optical Networks Eric Bouillet, Georgios Ellinas,
Jean-François Labourdette, Ramu Ramamurthy © 2007 John Wiley & Sons, Ltd